R00479 55602

CHICAGO PUBLIC LIBRARY
HAROLD WASHINGTON LIBRARY CENTER

R0047955602

D1519243

Lectures in
Theoretical Population Biology

We must learn from the mathematician to eliminate and to discard; to keep the type in mind and leave the single case, with all its accidents, alone; and to find in this sacrifice of what matters little and conservation of what matters much one of the peculiar excellences of the method of mathematics.

—*D'Arcy W. Thompson*
"On Growth and Form"
Cambridge University Press, 1917

Lectures in Theoretical Population Biology

Lev R. Ginzburg
State University of New York,
Stony Brook

Edward M. Golenberg
University of Haifa

Prentice-Hall, Inc., Englewood Cliffs, New Jersey 07632

Library of Congress Cataloging in Publication Data

GINZBURG, LEV R.
  Lectures in theoretical population biology.

  Includes index.
  1. Population biology. I. Golenberg, Edward M.,
(date). II. Title.
QH352.G556 1985   574.5'248   84-11586
ISBN 0-13-528043-5

Editorial/production supervision and
  interior design: *Kathleen M. Lafferty*
Cover design: *20/20 Services, Inc.*
Manufacturing buyer: *John B. Hall*

©1985 by Prentice-Hall, Inc., Englewood Cliffs, New Jersey 07632

*All rights reserved. No part of this book
may be reproduced, in any form or by any means,
without permission in writing from the publisher.*

Printed in the United States of America

10 9 8 7 6 5 4 3 2 1

ISBN 0-13-528043-5 01

Prentice-Hall International, Inc., *London*
Prentice-Hall of Australia Pty. Limited, *Sydney*
Editora Prentice-Hall do Brasil, Ltda., *Rio de Janeiro*
Prentice-Hall Canada Inc., *Toronto*
Prentice-Hall of India Private Limited, *New Delhi*
Prentice-Hall of Japan, Inc., *Tokyo*
Prentice-Hall of Southeast Asia Pte. Ltd., *Singapore*
Whitehall Books Limited, *Wellington, New Zealand*

Contents

To the instructor *vii*

To the student *ix*

Lecture 1   Introduction   *1*

I   Natural selection   7

Lecture 2	Hardy-Weinberg law	*7*
Lecture 3	Selection in natural populations	*15*
Lecture 4	Fisher fundamental theorem	*22*
Lecture 5	How selection works	*29*
	Problem solving	*37*
	Homework exercises	*44*

II   Other evolutionary forces   47

Lecture 6	Mutation	*47*
Lecture 7	Migration	*54*
Lecture 8	Inbreeding	*59*
	Problem solving	*70*
	Homework exercises	*77*

v

III Genetic drift and neutral evolution 81

Lecture 9	Sampling errors	81
Lecture 10	Drift in small populations	90
Lecture 11	Neutral evolution	96
	Problem solving	102
	Homework exercises	107
Lecture 12	Today's view of evolution	109

IV Population growth 117

Lecture 13	Malthusian growth	117
Lecture 14	Logistic growth	124
	Problem solving	132
	Homework exercises	140

V Interacting populations 143

Lecture 15	Lotka-Volterra competition models: competitive exclusion	143
Lecture 16	Lotka-Volterra competition models: coexistence	153
Lecture 17	Predation	160
Lecture 18	Ecosystem models	172
	Problem solving	180
	Homework exercises	190

VI Demography 193

Lecture 19	Leslie matrices	193
Lecture 20	Growth rate and equilibrium age distribution	201
Lecture 21	Demography applied	212
	Problem solving	220
	Homework exercises	232
Lecture 22	Ecological-genetical interactions	236

Index 243

To the instructor

Lectures in Theoretical Population Biology is an elementary undergraduate text. It has evolved out of the course, "Adaptation and Evolution," which has been given at the State University of New York at Stony Brook since 1977. The course attracted between 100 and 200 students per semester, these students majoring predominantly in biology, but also in physics, mathematics, and humanities. It was, for all the students, their first exposure to population biology covering areas of evolution and ecology, and for the majority of them it was also their last.

The first function of this course was, therefore, to present a broad, and necessarily sketchy, coverage of the field for the students who would never study the subject more deeply. For about 50 students, however, the course was the starting point of a more serious interest in the field. They went on to take other more specialized junior- and senior-level courses in the areas of ecology and evolution. A subset of these students continued working in the area and went on to graduate schools. The second function of this course was, therefore, to serve this "population biology inclined" group. The two groups are clearly very different and it is a challenge for the teacher to create a proper balance: not to lose the attention of the "terminal students" and not to fall into triviality, but always prodding and stimulating the more dear-to-the-heart, interested group. We submit the final result for your judgment to decide whether we were able to reach the proper balance.

The course was traditionally made up of two complementary parts: one more mathematically inclined and the other more field-oriented. This book presents the material taught in the theoretical portion of the course. We expect, therefore, that this text will be complemented by empirical natural history material.

We decided to organize the book in lectures numbered sequentially throughout. The subject of every lecture and the volume of material is designed to correspond to one lecture. We have tried to make each lecture a self-contained entity. References to the material of previous lectures are, of course, unavoidable, but we never make cross-references between formulas. If a particular formula is needed again, it is repeated as in an actual lecture.

We have included additional problem-solving material in the book. Given the limited time in class, this material can be used in recitation sessions. Every group of related problems is preceded by a short summary of information that is minimally necessary for solving these problems. Students find such summaries useful for test preparation.

We are greatly indebted to many people for their varied, though all indispensable contributions. Steve Peterson wrote most of the homework problems at the end of each chapter, and criticized and commented on our text throughout. Peter Frank, University of Oregon; Paul E. Hertz, Barnard College; Batia Lavie, University of Haifa; Robert May, Princeton University; and Robert B. Merritt, Smith College, reviewed the original manuscript and made many perceptive and helpful comments. Josef Ackerman and Matthew Liebman proofread the manuscript and made several useful suggestions on both style and substance. Shirley Felicetti typed the numerous drafts of the manuscript and tried, mostly in vain, to teach us some English grammar. Our wives, Tatyana and Zipora, added their own unique contributions through their criticism and encouragement. The quotation on page ii is used with permission from Cambridge University Press.

No textbook is in full agreement with the particular goals that a teacher may have, or with the particular students who happen to sign up for the course that semester. We have attempted to create a flexible structure for the course that can be easily filled out and enriched by every individual instructor in a a variety of ways to suit his or her particular circumstance.

Lev Ginzburg
Edward Golenberg

To the student

The words "first," "elementary," "beginning," and so on do not mean "superficial." Although this is your first encounter with the area of theoretical population biology, the things you will learn are serious science and the current subject of intense studies. The ideas that you will learn are directly related to many other areas, beginning with the widely discussed evolutionary controversies and ending with problems of environmental impact assessment.

The greatest enjoyment of learning is suddenly to see the connection between facts or phenomena that previously seemed unrelated. The number and extent of such connections that one recognizes has even been suggested as a measure of one's wisdom. We hope to establish a few new connections in your mind during this course. They will relate to theoretical population biology, an area of science which is going through the same process that physics went through three centuries ago. At that time, many laws of physics were qualitative, such as, "If you put hot and cold bodies together, the hot one will cool down and the cold one will warm up." You may laugh at this sort of naiveté, but biology, to a large extent, is still at this level of description. It is not completely senseless; it does give some idea of what is going on. However, we are not satisfied with such descriptions anymore. We know that it was possible to give a much fuller quantitative picture of how heat is transferred. Likewise, we hope to come to the same kind of understanding in other areas of science.

To encourage this kind of understanding in biology, we can develop mathematical models that are quantitative descriptions of the processes which occur in biological populations. This book is about models. Very few of them are called laws. The difference between a law and a model is only qualitative. Important, widely applicable models are promoted into the status of laws. Both, however, are always approximate formalized descriptions of infinitely complex realities. Some are just superior to others. Only time can show which models will evolve sufficiently to the rank of laws.

This transitory character makes it particularly exciting to work in the field. One can hope to make a greater contribution, to discover something that is really important, in an area which is still young and not fully stabilized. We hope you will find the subject interesting, even if you are not planning to work in this field and are taking the course out of general curiosity or simply to satisfy a formal requirement. It does not make much sense today to study seventeenth century physics unless you are interested in the history of science, and very few of you are. By the same token, if you are going to study population biology, you should do it on the modern level from the very beginning. This is what this course is designed to provide for you.

A knowledge of genetics on the level of an introductory general biology course is required to understand the material in this course. One year of calculus is another requirement. Even if you are more knowledgeable in genetics or in mathematics than minimally required, you will probably not find any repetition of what you already know in this book. You will just see more than others and you will make more useful connections.

Lev Ginzburg
Edward Golenberg

lecture 1

Introduction

The development of science is driven by intuition. You may be used to hearing about the wonderful structure of science. One fact follows beautifully and necessarily from the other; there are perfect, logical sequences of conclusions, and basic facts are experimentally verified to be correct to the sixth decimal place. If this were true, what would be the stimulus for new developments and how would one generate new knowledge? Does new knowledge come about by logical deduction from previously collected knowledge? Let us consider, for instance, mathematics. It is supposed to be even purer than natural sciences. Here, some may say, everything is absolutely clear and theorems are just the logical consequences of axioms. The tree of all possible theorems makes what we call mathematics. If this were the case, how do new axioms appear? They cannot be logically deduced from previous axioms. In another example mathematician A decided to prove theorem X, worked on it for a year, and finally proved it to be correct. What made him decide to prove this particular theorem rather than some other theorem? He did not have the logical proof when he started the work.

Intuition is the only answer. Intuition is always our guide and master in extending knowledge. We basically guess the result and only then try to prove it. If we are able to prove that it is incorrect, we try another guess. Although the process of checking out the accuracy of the guess is completely logical, the original process of formulating what to check is intuitive and relatively unclear. If this is not convincing enough, why is the profession of mathematics not eliminated by computers? I believe that the computer is, in principle, capable of finding all the logical consequences of a *finite* set of axioms with a *finite* set of deductive rules. However, there are serious doubts about the computer's ability to generate new axioms. Are not laws of natural sciences similar to axioms?

My advice is to train your intuition. If your intuition leads not perfectly, but at least close to correct conclusions, then from time to time, you are going to guess right. No one has perfectly correct intuition. Some people have it better in one field than in others. What is important is that intuition is trainable. You can greatly improve it by learning the subject in which you would like to be able to be intuitive. I also believe that there is a strong relationship between your intuitive abilities in different areas. Once you have trained yourself in one direction, you will be able to learn and understand concepts in other directions much faster.

It is a common belief that a heavy body will fall to the ground faster than a light body. (We assume equal shapes of bodies to equalize air resistance and equal heights from which they fall.) This incorrect intuition has dominated minds for close to 2000 years. Galileo was the first to prove, logically and experimentally, that bodies will fall in equal time independently of their weight. His proof is faulty, but absolutely marvelous. Let me defer discussing Galileo's proof until the last lecture. What is important to stress is that even today, centuries after Galileo's discovery, many people still have the wrong intuition. It is one thing to substitute numbers in the formula and see that answers come out the same for both bodies. It is another to see it intuitively.

When you understand a subject, it should become part of your consciousness and then you can jump toward the unknown, you can make a wild guess. Let the logical part of your brain, or your time in the laboratory, or both, confirm whether or not it was a good jump. Without the jump, nothing happens.

The mathematical models in population biology, as well as the physical laws, will help to change your intuition, if it were wrong before, or create an intuition in an area where you did not have any. After we develop a particular model, we will look at all its consequences. Sometimes we will encounter statements that seem "counterintuitive." These are the challenging statements: Either the model was completely wrong or our intuition is faulty and should be retuned. In most instances the latter is the case.

Imagine that we have two large populations of fruit flies (*Drosophila*). Both are samples brought to the laboratory from nature. Some of the flies have red eyes, others have white eyes. The difference in the percentage of red-eyed flies in the two samples is only a few percent. The two samples are then placed in identical population cages and cultured in identical conditions for many generations (the generation time of fruit flies is about 2 weeks). After a long time (say, a year), one population becomes 100% red-eyed and another is 100% white-eyed—completely opposite results. The environments were the same and the flies were basically the same except for a few percent difference in the frequency of this trait noticed at the beginning. Intuition asserts that we are dealing with the natural selection process. Yet for some reason, in one cage the red-eyed fruit fly was more fit and in the other cage the white-eyed fruit fly was more fit. This cannot be right: The environments were *identical*. The same type cannot be more fit in one cage and less fit in another; the fitnesses should be the same. Suppose that the red-eyed type is always more fit. If I understood this experiment correctly, Charles Darwin was wrong. In one of the cases the type that was less fit actually won and a more fit type disappeared. Darwin's idea of natural selection seems to be correct, why didn't it work in this case?

This experiment is realistic; there is nothing wrong with it. Yet, at the same time, Darwin was also correct, in principle. However, he did not know one important thing at the time of his discovery: He did not know genetics. What does this change? Actually, it changes our understanding completely. Let me postpone the answer to this puzzle. In about three lectures you will have a good, plausible hypothesis of why we could have observed such a result.

Life would be too easy if this were the end of the story. In about eight or nine lectures we will develop an alternative theory which might also explain the result. The second explanation, completely non-Darwinian, will be based on the neutral theory of evolution which is an extremely interesting development of the past 20 to 25 years. Having more than one explanation is, to a certain extent, as bad as not having one. We are left with the problem of distinguishing between them, and this, in itself, is not a trivial problem. The goal of this introduction is, however, to give a general idea of the course. The problem which we have just discussed may give an impression of the type of problems we will encounter.

I also mentioned that some models are superior to others. How does one distinguish a superior model from an inferior one? There are so many sides to this issue that there is a risk of becoming too philosophical before we have begun to learn anything substantial. However, there is one important general way of comparing models which should be discussed from the beginning.

As an example, recall that the trajectories of planets around the sun are approximate ellipses and that this follows from Isaac Newton's laws. However, people were able to predict the locations of all visible planets very accurately long before Newton. The widely accepted theory was the Ptolemaic model. It works extremely well, and at the time of Newton it was more accurate than the new theory. The idea is as follows: Earth is in the center and the movement of any planet is fitted by a simple rotation around Earth. The sun and moon work fairly well because their actual ellipses are not too far from circles. If you are not satisfied with the accuracy, imagine another point which rotates around the first, again with constant angular speed. If this is not enough, make the third point rotating round the second, and so on. By fitting appropriate radii of the circles and angular speeds, one can fit any *periodic* movement in the sky, however complicated, as seen from Earth. You do realize that, if we look from Earth, the planets of the solar system follow very complicated trajectories. They are repetitive, though, and this is all that is required for the Ptolemaic system to work. This was a successful and general model and it gave quite accurate numerical predictions about the future positions of planets. The fact that the Earth is at the center does not detract from the elegance of the model.

We now know the laws of Newton which tell us what the trajectory of a planet should be. Somehow, people immediately recognized the superiority of Newton's model over Ptolemy's. If both describe and predict the future equally well, what is the fundamental difference?

The difference is in the rejection power. Ptolemy's system does describe what happens but it will also find radii and angular speeds for any elliptical curve that can be arbitrarily drawn in the sky. If you make up your own imaginary planet with any trajectory you want, as long as it is periodic, and submit your "findings" to Ptolemy, he will not be surprised. He will carefully find all the necessary parameters and your imaginary trajectory will be accurately predicted for all its future repetitions. If you bring your "planet" to the modern physicist, he will immediately say that you are lying; there is no such planet and it cannot exist because the trajectory contradicts the basic laws of physics. Thus the new theory describes what can happen and only that, rejecting a multitude of other movements as impossible, whereas the old model describes, but does not reject. The stronger the rejection power, the more powerful the theory.

Let us examine our present understanding of evolution from this point of view. What if we now invent a nonexistent animal or plant, but make it reasonably believable: We describe its environment, life history, and appearance, draw it, and tell the evolutionary biologist about our great new discovery. Will he be able to say, "No, this is impossible; this is against evolutionary laws. There cannot be such a creature on Earth. You made it up." In many cases, unfortunately, our hypothetical scientist will not be able to reject the invented species. There was actually a man who made stuffed models of nonexisting birds by altering real specimens and successfully sold them to respectable museums! Does this mean that evolutionary theory is still in the Ptolemaic era, that we do not have a decisive and clear understanding of the evolutionary process to a degree similar to that of planetary astronomy? Yes, it does. There are, however, things we can reject even today. A creature that looks like a fish with fins and scales is not going to live in a tree—this much we know. We would like to know much more. We would like to be able to describe the actual evolutionary process in clear terms. We want both qualitative and quantitative understanding of what is going on in living populations. The mathematical models that we will begin to study in this course are aimed at providing such an understanding.

It is common to hear that biology is a much more complex subject than physics, that nonliving bodies are somehow easier to describe than live organisms. It might be correct, but those who say that probably do not realize how diverse the movements of inanimate objects are, and what a variety of shapes, colors, smells, and other traits they have. It is only the genius of the founders of physics that clarified that mass is all that is basically important for the laws of mechanics to explain a universe of trajectories in a concise and clear manner. I would rather believe that, being overwhelmed by the diversity and seeming complexity of living things, we have not found the right point of view or the position at the "top of the mountain" from which all will suddenly be clarified and fall into place. It is a great challenge to come up with such a point of view. One reason that biology, and evolutionary theory in particular, is such a difficult subject is its historical character. As much as for the history of human society, the problem is that we cannot repeat the experiment. We are not able to check what would have happened if Napoleon had decided not to invade Russia as he was advised. Again, the believers in the plague of historical influences in biology ignore the fact that, until fairly recently, physics was as "historical" as biology. The point of view that "forces produce speeds," and "if nothing pushes it, no movement will result," prevailed from Aristotle until the eighteenth century. The Newtonian revolution, relating force to acceleration rather than speed, was a point of view which made physical laws nonhistorical. The new theory, however, required initial positions and speeds to be given from outside the theory. Physics does not answer the question of why the earth has, or ever had, a particular position and speed relative to the sun. This is the subject of cosmology, a historical science with all its schools of thought and controversial issues. We are left in physics with clear, nonhistoric laws, but physics is defined differently than it was previously. The "cosmological" questions concerning the history of the solar system were a part of physics; questions about positions and speeds seemed to be quite reasonable. Now we say that they are "initial conditions" and real laws describe what happens, given these conditions, but do not describe or explain the conditions themselves. Some believe that even the law of gravity is only approximately nonhistoric. The gravity may not really be constant. It may change slowly as a function of the distribution of mass in the universe. Thus nothing is exact in science and reasonably good approximations are all that we can hope for.

Perhaps we are asking questions that are too hard to answer with our modern evolutionary theory. What if we narrow the set of questions relegating some of the parameters to the "initial conditions" as physicists do? Can we set aside all the history-dependent questions and ask only the questions that are answerable in an approximately history-independent way? We are trying to do this using our mathematical models. The predictions that arise from the models are based on a number of initial conditions. The model predicts future evolution only with given initial states. In other words, disregarding history, if we are in a given position, what is the next move? Are there any interesting questions left within this nonhistoric subset or are all the interesting questions in biology fundamentally historic? Does this contrast biology with physics?

One of the definitions of a pessimist is that he is just a well-informed optimist. I may not be well enough informed and I continue to learn. At the moment I am still an optimist. I will attempt to project some of the reasons for my optimism in the following lectures.

ခhapter 1

Natural Selection

lecture 2

Hardy-Weinberg law

This chapter, along with the next two, will introduce the subject of *population genetics*. In a nutshell we can describe population genetics as that which concerns itself with the mechanisms that drive evolution. So let us begin and ask what is evolution?

When you think of evolution, you may conjure up drawings of Mesozoic habitats with images of dinosaurs, gigantic cockroaches, and flowering magnolias. The dinosaurs are now extinct, the roaches are smaller, and the magnolias are pretty much the same. If I were to ask which organism had undergone evolution, you would probably respond that the cockroaches did. They are the ones that are still around, but have in some way changed. Evolving implies changes in traits of organisms.

Evolution may be defined as the changes in traits, but which traits? Long, bushy sideburns in men seem to come into fashion with a periodicity of about 80 years. If we would collect pictures of everyone's great-grandfather, grandfather, and father and notice the degree of facial hair, we would see a general trend in hirsuteness. Is this evolution? No, at least not in the biological sense. The types of traits that we are interested in as evolutionary biologists are those that are determined by the genetic material of the organism, that is, those traits that are heritable.

Finally, it should be clear that we cannot be talking about changes in heritable traits in individuals. With the possible exception of somatic mutations in some plants, a change in a trait in an individual organism is not usually caused by a change in the genotype of that organism. When a person's hair turns white with age, it is not because the gene controlling hair pigmentation has changed. There is simply a developmental change occurring in that person.

When we look at changes in genetic characters, we look at *collections* of interbreeding individuals, or *populations*, not individuals. What is meant by a population? First, the commonest misconception is that we are talking about human populations. Many things have populations, not only humans. You can even talk about a population of inanimate objects, using population in a loose sense. When I use the term *population*, however, I am referring to a collection of interbreeding organisms. They could be humans, seals, robins, mackerel, violets, or bacteria. Second, the population may consist of all the members of a species or a smaller subgroup of that species.

If we are not following the change of occurrence of a trait in one individual but rather in a collection of individuals, then we are interested in the average occurrence of this trait. This average occurrence is called the *frequency* of the trait. We can define frequency in the following way:

$$p = \frac{\text{number of occurrences of a trait}}{\text{sample size}}$$

For example, suppose that I want to find the frequency of the Y chromosome in the population of students signed up for this course. Let us say that there are 102 signed up and 49 of these are males. Every individual has two sex chromosomes so that the denominator is $204 = 2 \times 102$. The number of Y chromosomes can easily be determined by counting the number of males in the course, since each male has one Y chromosome. The frequency can then be determined: $p = 49/204 = 0.24$.

A frequency can be thought of as a *probability* of selecting that trait if you could blindfold yourself and sample from the population. Looking at our Y chromosome, the frequency is equal to 0.24. Assuming that each individual produces an equal number of gametes, the probability of selecting a gamete carrying a Y chromosome is 0.24. This is a basic concept, but it is essential in order to understand the rest of our discussions.

Now then, back to our definition of evolution. Let us use the symbol Δ to mean the change of something over a discrete period of time. In our case this period of time is one generation. We may semiformally write our definition of evolution as Δp, that is, evolution is the change of frequency of a heritable trait over time. When Δp is equal to zero, there is no evolution of that trait;

Lecture 2   Hardy-Weinberg Law

when Δp is not equal to zero, the population is evolving with respect to that trait. If the change is positive, the trait is becoming more prevalent in a population over time. If it is negative, the trait is becoming rarer. This type of evolution—the changes in frequency of genetic traits—is called *microevolution*. Microevolutionary changes are commonly thought of as the building blocks to greater physiological, morphological, or ecological changes of a population or species. These changes, over long periods of time, constitute *macroevolution*. In population genetics we attempt to understand evolution by analyzing the building blocks. I should mention that this is not a universally accepted point of view of the relationship between microevolution and macroevolution. I will discuss this controversy further in Lecture 12.

Let us start formally to build our models to describe microevolution. In doing so we will try to increase the rigor of our thinking and to expand on or correct the way our intuition tells us that evolution works. We will look at one *gene locus* in a population of *diploid* organisms that has *two alleles*, a and b. We will explicitly state the following five assumptions:

1. The population is large.
2. There is no mutation.
3. There is no natural selection.
4. There is no migration.
5. There is panmixia, that is, all individuals have an equal chance of mating with each other.

What will be the frequency of the a and b alleles after one generation? How will the diploid genotypes be distributed after one generation?

The frequency of allele a will be defined as $f_a = p$ and the frequency of allele b as $f_b = q$. I know that $p + q = 1$ because there are only two alleles. I also know that there will be *three genotypes* in the population for this gene locus—aa, ab, and bb. I do not yet know the frequency of the aa genotype, but I will assign it as $f_{aa} = P$. Similarly, I will call the frequency of ab genotypes $f_{ab} = Q$ and the frequency of the bb genotype, $f_{bb} = R$. I know that $P + Q + R = 1$ because these are the only genotypes. I also know from our definition of frequency that p = (number of a alleles in genotype aa × the frequency of genotype aa + number of a alleles in genotype ab × the frequency of genotype ab + number of a alleles in genotype bb × the frequency of genotype bb)/(number of all alleles in all genotypes). This is then

$$p = (2 \times P + 1 \times Q + 0 \times R)/(2 \times P + 2 \times Q + 2 \times R)$$
$$= (2P + Q)/2(P + Q + R)$$

But $P + Q + R = 1$, so $p = P + \frac{1}{2}Q$. By similar reasoning $q = \frac{1}{2}Q + R$.

What happens after one generation? The probability that a particular pair of genotypes will mate in a panmictic population is simply the probability of them meeting. This is identical to the probability of sampling two individuals of the particular genotypes. Since this in turn is simply two independent sampling events, the probability of the occurrence of a given pair of genotypes is simply the multiple of their individual probabilities or frequencies in the population. For example if P is the frequency of an aa genotype in the population, then the probability of a pair of two aa genotypes is $P \times P$. Furthermore, we can predict the frequencies of the genotypes of the offspring of each individual mating by applying Mendelian genetics. For example, when two aa individuals mate, we know that all the offspring will be aa. The following table includes all the possible mating pairs, their probability of occurrence, and the frequencies of their offsprings' genotypes. From this table I can determine the frequencies of each allele and genotype in the next generation.

		Offspring Frequencies		
Pairs	Probability	aa	ab	bb
$aa \times aa$	$P \times P$	1		
$aa \times ab$	$P \times Q$	1/2	1/2	
$ab \times aa$	$Q \times P$	1/2	1/2	
$ab \times ab$	$Q \times Q$	1/4	1/2	1/4
$ab \times bb$	$Q \times R$		1/2	1/2
$bb \times ab$	$R \times Q$		1/2	1/2
$bb \times aa$	$R \times P$		1	
$aa \times bb$	$P \times R$		1	
$bb \times bb$	$R \times R$			1

The frequency of each genotype in the next generation, which we will denote by the prime ', is

$$f'_{aa} = P^2 + \frac{2}{2}PQ + \frac{1}{4}Q^2$$

$$= (P + \frac{1}{2}Q)^2$$

but

$$P + \frac{1}{2}Q = p$$

Hence,

$$f'_{aa} = p^2$$

$$f'_{ab} = \frac{2}{2}PQ + \frac{1}{2}Q^2 + \frac{2}{2}QR + 2PR$$

$$= 2\left[\frac{1}{2}PQ + \frac{1}{4}Q^2 + \frac{1}{2}QR + PR\right]$$

$$= 2\left(P + \frac{1}{2}Q\right)\left(R + \frac{1}{2}Q\right)$$

$$= 2pq$$

$$f'_{bb} = \frac{1}{4}Q^2 + \frac{2}{2}QR + R^2$$

$$= \left(R + \frac{1}{2}Q\right)^2 = q^2$$

So we see that no matter what the original distribution of genotypes, the population will go to the distribution of p^2, $2pq$, and q^2 for the aa, ab, and bb genotypes, respectively, after only one generation of random mating.

Another way of considering this phenomenon is by analyzing the life cycle of organisms with nonoverlapping generations. We can present a simplified life cycle (Figure 2.1) as a sequence of stages consisting of adult diploid individuals, their haploid gametes, the newly created zygotes after fertilization, and then back to mature diploid individuals. I have presented the problem by starting with adult diploid individuals. The actual stage which randomizes our genotype distribution and leads to the genotypic frequencies of p^2, $2pq$, and q^2 is the formation of the new zygotes out of the collection or *pool* of gametes. If we start our analysis at the haploid stage, we do not need to wait one generation for this to be established. However, we usually begin natural or laboratory studies of populations with mature individuals, not their gametes, so that one generation of random mating is required.

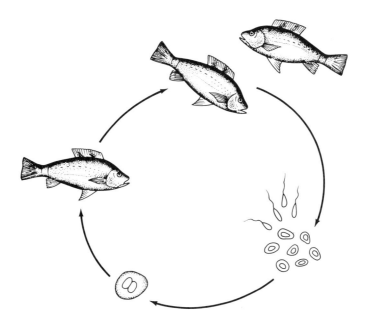

Figure 2.1 Schematic presentation of a simple life cycle indicating summary stages where selection may occur.

What are the new frequencies of the a and b alleles, p' and q'?

$$p' = f'_{aa} + \frac{1}{2}f'_{ab} \qquad \text{and} \qquad q' = f'_{bb} + \frac{1}{2}f'_{ab}$$

$$= p^2 + \frac{1}{2}(2pq) \qquad\qquad\qquad = q^2 + \frac{1}{2}(2pq)$$

$$= p(p+q) \qquad\qquad\qquad\qquad = q(q+p)$$

$$= p \qquad\qquad\qquad\qquad\qquad\quad = q$$

So $\Delta p = p' - p = p - p = 0$. There is no change in frequency given the assumptions stated.

Try an example to test this first finding. An extreme case will be when there are no heterozygotes, so that $P = p$, $Q = 0$, and $R = q$. Our table collapses to

		aa	ab	bb
aa × aa	P^2	1		
aa × bb	PR		1	
bb × aa	PR		1	
bb × bb	R^2			1

$$f'_{aa} = P^2 = p^2 \qquad f'_{ab} = 2PR = 2pq \qquad f'_{bb} = R^2 = q^2$$

You may argue that there are not many real populations of organisms that fit these assumptions. Assumptions in models generally attempt to simplify the real world so that we may understand the mechanisms behind phenomena. Obviously, then, no real situation will exactly fit the assumptions of a model. Indeed, the more unrealistic the assumptions, the less useful the model. After all we have done, is our model quite useless? The following example is from a real population, one which you will swear does not fit the assumptions. See how we use the model to fit real data.

These data are from a study of frequencies of M-N blood groups in a sample of 1279 people. The observed frequencies are

Genotype	MM	MN	NN
Frequency	0.2838	0.4957	0.2205

We first have to calculate the observed allelic frequencies.

$$f_m = p = f_{MM} + \frac{1}{2}f_{MN}$$

$$= 0.2838 + \frac{1}{2}(0.4957) = 0.2838 + 0.2478$$

$$= 0.5316$$

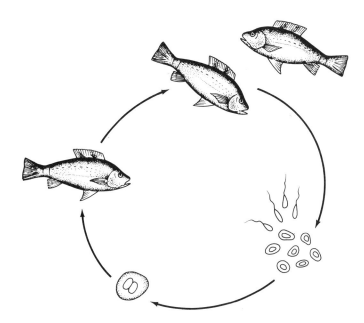

$$f_n = q = 1 - f_m = 0.4684$$

We calculate the expected zygotic frequencies as follows:

$$f_{MM} = p^2 = 0.2826 \qquad f_{MN} = 2pq = 0.4980 \qquad f_{NN} = q^2 = 0.2194$$

If we were to make a statistical test of the closeness of fit of the expected to the observed values, we would find that there is no significant difference between the two. Why? What about the assumptions that we made?

The first assumption may be true. The population is fairly large and this is only a sample of the larger population. Is there mutation? Probably, but the chance of a mutation in the germ line from one allele to another specific allele is small. We will see shortly that the expected deviation in one generation due to mutation is not great from this effect. There is natural selection on human populations even though it has been reduced considerably. There definitely is migration in human populations. Finally, there is certainly no panmixia in human populations. Social and cultural considerations are usually of the utmost importance in mate selection. Why *do* the data support our expectations?

The answer lies in the fact that all these assumptions are made in regard to one particular locus. If immigration and emigration are random with regard to the *M-N* locus, then the assumption is not violated. If there is no selection on that particular locus, or loci linked to it, then whether or not there is selection on the organism is also irrelevant, Finally, for all the myriad considerations we have in our mate selection, *M-N* blood groups are not among them.

Let us review the essential points. First under the five assumptions that we made concerning our population, the frequency of an allele will not change from one generation to the next. This is true no matter what the frequencies are in a population. This is analogous to a ball on a flat surface. Wherever we place that ball it will remain there; it will not roll in any direction no matter where it is located. The allelic frequencies are, therefore, said to be in *equilibrium*. Since the same population will remain in equilibrium, regardless of its frequencies, we call this a *neutral equilibrium*.

Second, the frequency of a particular genotype may change after one generation given our assumptions. However, after only one generation, the frequencies of our genotypes will reach the frequencies p^2, $2pq$, and q^2 and will remain at these frequencies in all future generations. These frequencies are called the Hardy-Weinberg equilibrium frequencies, in honor of G. H. Hardy and V. W. Weinberg, who derived the formulas at the beginning of this century.

We may extend the Hardy-Weinberg concepts to more than two alleles. For example, if we have three alleles *a*, *b*, and *c*, we can assign their frequencies to be *p*, *q*, and *r*, respectively. The frequencies of the homozygous genotypes *aa*, *bb*, and *cc* will be p^2, q^2, and r^2 after one generation. The frequencies of the heterozygotes *ab*, *ac*, and *bc* will be $2pq$, $2pr$, and $2qr$. You can expand this to any number of alleles by noting that the expected frequencies are derived from the terms of the quadratic expression $(p + q + r + \ldots)^2$. We will practice using the Hardy-Weinberg frequencies with three alleles in a problem at the end of the chapter.

The Hardy-Weinberg frequencies, and the conclusions which we derived from them, make up our first *law* of population genetics. Yet it is so simple. Mathematically, it was known long in advance of Hardy and Weinberg. If they had not "discovered" it, you could probably derive the same law today. It is simply the result of expanding Mendelian laws to the population level, seeing a new generation of zygotes as a repeated sampling of alleles or balls from an urn. Indeed, many others independently came to the same conclusion, and you will often see names such as W. Castle and S. S. Tschetverikov listed with Hardy and Weinberg. Of the two, Weinberg was a biologist whose other work is not well known. Hardy, however, was an eminent English mathematician. It is ironic that he is most widely known for such a mathematically trivial equation. It is not the equation for which we give recognition, but rather for the scientific intuition that applied it to genetics.

lecture 3
Selection in natural populations

Using the model which we introduced in Lecture 2, we came to two conclusions: (1) The gene frequencies will not change and (2) the genotype frequencies will reach equilibrium values after one generation and will not change thereafter. Going back to our original definition of evolution, Δp, we see that our population is not evolving. It is continuing on through time maintaining its gene frequencies. This is analogous to a body in motion maintaining its velocity in the absence of applied force. Let us use some evolutionary force and see what will happen to our population.

The first force we will use is *selection*. When you think of selection, the first thing that probably comes to mind is "survival of the fittest." Yes, but who or what are the fittest? If you answer, "The fittest are those that survive," you are cautious but not very informative.

We might get a little further if we return to our original discussion of evolution. We said that we are interested in the change of frequency of traits in a population over time. Later, in discussing the Hardy-Weinberg equilibrium, we touched briefly on the concept of disembodying the gene from the organism in which it acts. Let us put these two concepts together in our discussion of selection and survival of the fittest. The thing that is fittest is now the allele and it will persist or survive over time. An allele can persist indefinitely if it is passed on from one organism to the next via a gamete. Even though an organism dies, a copy of the allele may be transmitted to, and thus perpetuated through, its offspring. Differences in survival or perpetuation between alleles will occur if the alleles are represented in the offspring generation in a different proportion or frequency from that of the parental generation. If there is a difference in frequencies between the two generations, then Δp is not equal to zero and there is evolution. If the difference is caused by particular systematic forces, then we can say that there is selection occurring.

To what particular systematic forces are we referring when we talk about selection? These are forces that act on an organism during its life cycle and either reduce or increase its probable number of offspring relative to other individuals in the population. With very little effort, we can think of dozens of such forces in only a few minutes. For example, if I were thinking of a seed plant, some of these forces would be the amount of endosperm available to the growing seedling, the coordination of germination with the last frost of spring, the amount of water transpired through the stomata during the day, the aggressiveness of the roots in spreading, the quantity or quality of compounds to discourage herbivores, the size of the flower, the amount of nectar to attract pollinators, the timing of flowering to increase the number of coincident flowering neighbors, the amount of photosynthate shunted to the developing seeds, and the mechanism of dispersal. It should be obvious that the more we know about a particular plant or animal, the more factors we can think of which will significantly affect the probable number of offspring. It should be even more obvious that to enter each specific force as a factor in our model would be exceedingly complicated. We can mentally divide an organism's life cycle into three elements: growth and maintenance, mate selection, and reproduction. All these individual factors can be summarized into one of three groups or components affecting the fitness according to the part of the life cycle in which they occur. Factors affecting growth and maintenance are grouped under the viability component. Those factors affecting mate selection are considered under the mating success component. The fecundity component encompasses the actual number of offspring produced per mating. Taken together, these separate elements or components constitute the *fitness* of an individual.

Let me clarify a central point that was made earlier. We talked about the survival of the fittest allele. We must modify that concept slightly to add biological realism. All the systematic forces given previously act directly on whole organisms of a given genotype. Although we can talk about the fitness of an allele, that fitness is determined by averaging the fitnesses of all the genotypes containing this allele.

What is of direct interest is the fitness of diploid genotypes. The *absolute fitness* of a genotype is the average number of viable offspring it produces. However, in evolution we are not interested in absolute numbers, but in relative frequencies. Similarly, we are interested in the relative number of viable offspring produced. If a genotype produces twice as many offspring as another, it makes no difference whether the actual numbers are 2 and 1 or 50 and 25. The relative effect is the same. Therefore, as a convention and to make life easier, we set the highest fitness to one and all the others are given as proportions of the highest fitness. These are called *relative fitnesses*. For example, say we have a population of flowering plants which has three color morphs, red, pink, and white, and these morphs are controlled by two alleles, R and r, at one locus. The red morph is RR, the pink Rr, and the white rr. If the red flowering plant leaves 25 seedlings on the average, the white 30, and the pink 50, then their *absolute fitnesses* are 25, 30, and 50, respectively. Their *relative fitnesses* are 0.5, 0.6, and 1, respectively. By convention, relative fitnesses of a genotype are designated by the letter W with a subscript for the genotype. For example, in the preceding case W_{RR} is equal to 0.5, W_{Rr} is equal to 1.0, and W_{rr} is equal to 0.6.

Let us take our equilibrium model from the last lecture and add the effects of different relative fitnesses. In effect we are maintaining the same model but are just releasing the third assumption: We are allowing selection to act on our system. As you will recall, we defined the frequency of the allele a in a diallelic (two-allele) system as p and its frequency after one generation, p', as $p' = (p^2 + pq)/(p^2 + 2pq + q^2)$. This was determined by the average number of a alleles produced by each genotype multiplied by the frequency of that genotype divided by the total number of all alleles produced. This, we demonstrated, reduces to $p' = p$ and thus $\Delta p = 0$.

To accommodate our model of selection, all we have to do is to take into account that each genotype will not produce an equal number of alleles. Whereas before for each term in the numerator we multiplied the number of a alleles by the frequency of each genotype and assumed that each genotype produced the same total number of alleles, now we must add an extra weighting term for different relative fitnesses. For example, previously we stated that the genotype aa will produce two a alleles for every two alleles any other genotype may produce. The contribution of this genotype to the total number of alleles is weighted by its frequency in the population. The term in the numerator representing the contribution of a alleles by aa individuals is $2p^2$. Take an extreme case and say that, on the average, only 1 in 10 aa individuals reaches reproductive age compared to the other genotypes. It is obvious that we must take this into consideration when we calculate the contribution of the aa genotype to the next generation. We do this by multiplying our previous aa contribution by 0.1, so that the term is now $(0.1)(2)(p^2)$.

The fitness of the aa genotype, or W_{aa} by our notation, is 0.1. We have multiplied the contribution of this genotype term by our fitness constant. Let us now generalize and write the full equation for the frequencies of the two alleles a and b.

$$p' = \frac{W_{aa}p^2 + W_{ab}pq}{W_{aa}p^2 + W_{ab}2pq + W_{bb}q^2}$$

$$q' = \frac{W_{ab}pq + W_{bb}q^2}{W_{aa}p^2 + W_{ab}2pq + W_{bb}q^2}$$

These formidable looking equations are quite simple: Each individual term is multiplied by its relative contribution to the next generation, that is, its fitness. Unfortunately, the denominator no longer is equal to one. However, we can simplify our notation by noticing that our denominator is the sum of genotypic frequencies weighted by their fitnesses. In other words the denominator is equal to the *average fitness* of the population. This we may denote by the symbol $\bar{W}$. We may rewrite the equations as

$$p' = \frac{W_{aa}p^2 + W_{ab}pq}{\bar{W}}$$

$$q' = \frac{W_{ab}pq + W_{bb}q^2}{\bar{W}}$$

You should notice that $p' + q' = 1$. Our new frequencies still account for all the alleles.

You may ask if $p' = p$. We may restate the question by asking whether there is evolution under selection, or whether Δp does not equal zero.

We begin by restating the following definition:

$$\Delta p = p' - p$$

Let us now substitute our value for p' and then rearrange our equation.

$$\Delta p = \frac{W_{aa}p^2 + W_{ab}pq}{\bar{W}} - p$$

$$= \frac{W_{aa}p^2 + W_{ab}pq - \bar{W}p}{\bar{W}}$$

$$= \frac{p}{\bar{W}}(W_{aa}p + W_{ab}q - \bar{W})$$

Recall that $\bar{W} = W_{aa}p^2 + W_{ab}2pq + W_{bb}q^2$

Then $\Delta p = \frac{p}{\bar{W}}(W_{aa}p + W_{ab}q - W_{aa}p^2 - W_{ab}2pq - W_{bb}q^2)$

$$= \frac{p}{\bar{W}}[W_{aa}p(1 - p) + W_{ab}q(1 - 2p) + W_{bb}(-q^2)]$$

We know that $1 - p = q$, so

$$\Delta p = \frac{p(1 - p)}{\bar{W}}[W_{aa}p + W_{ab}(1 - 2p) - W_{bb}(1 - p)]$$

$$= \frac{p(1 - p)}{\bar{W}}[(W_{ab} - W_{bb}) + p(W_{aa} + W_{bb} - 2W_{ab})]$$

That was a lot of rearranging to get to an equation that does not look very polished. It does, however, tell us quite a bit. We see first that Δp will be proportional to the frequency of the *a* and *b* alleles, p and $(1 - p)$. If either allele is very rare, Δp will be small. This makes good, biologically intuitive

Lecture 3   Selection in Natural Populations

sense. If the frequency of an allele is small, few individuals in the population carry it, so that the effect on the population coming from individuals within the population is small.

Second, Δp will be proportional to the intensity of selection. The way that we have arranged the equation makes this point only subtly. Let us return to the second step in our manipulation of the equation for Δp.

$$\Delta p = \frac{W_{aa}p^2 + W_{ab}pq - \bar{W}p}{\bar{W}}$$

If I rearrange the equation such that

$$\Delta p = \frac{p}{\bar{W}}(W_{aa}p + W_{ab}q - \bar{W})$$

we can see that within the parentheses there are actually two mean fitnesses: $\bar{W}$ which is the population mean fitness and $(W_{aa}p + W_{ab}q)$ which is the mean fitness or *marginal fitness* of the a allele. An a allele can be paired to a or b in a genotype. The chances that the second one is a or b are correspondingly p and q. This expression is, therefore, the expected fitness of an allele averaged over the unknown partner in a zygote. We denote this by W_a. Our equation now simplifies to

$$\Delta p = \frac{p}{\bar{W}}(W_a - \bar{W})$$

Similarly, we can rearrange our equation for Δq such that

$$\Delta q = \frac{q}{\bar{W}}(W_b - \bar{W})$$

You can check these equations by showing that

$$\Delta p + \Delta q = \frac{1}{\bar{W}}[p(W_a - \bar{W}) + q(W_b - \bar{W})] = 0$$

One allele will increase in frequency at the expense of the other, $\Delta p = -\Delta q$. Thus if there is selection either for or against an allele in the population, the marginal fitness of that allele will be either higher or lower than the average fitness in the population. The stronger the selection pressure, the greater the difference. The greater the difference in these two measures of fitness, the greater the change in frequency Δp for a given generation. This agrees with our intuition of how selection should work.

Another way of seeing this clearly is to use the *coefficient of selection, s*, where s is the deficiency in fitness relative to the most fit genotype. If, for instance,

$$W_{aa} = 1 \qquad W_{ab} = 1 \qquad W_{bb} = 1 - s$$

we have

$$\Delta p = \frac{spq^2}{1 - sq^2} = spq^2 + s^2pq^4 + \cdots$$

If s is small, that is, selection against the homozygous bb individuals is weak, s^2, s^3, and higher powers of s are very small as compared to s. Therefore, we can simplify by disregarding these higher-order terms in s. We obtain

$$\Delta p \approx spq^2$$

We seem to have lost sight of our question: Does $\Delta p = 0$ under selection? By looking at our equation for Δp, we can see that there is not an obvious answer. It depends on the values of p, W_{aa}, W_{ab}, and W_{bb}. If W_{aa}, W_{ab}, and W_{bb} are all equal, then the model is reduced back to the Hardy-Weinberg equilibrium and there is no evolution. Even if there are fitness differences, there may be values of p at which there is no evolution. We may solve the equation when Δp is zero:

$$\Delta p = 0 = \frac{p(1-p)}{\bar{W}}[(W_{ab} - W_{bb}) + p(W_{aa} + W_{bb} - 2W_{ab})]$$

Looking at this equation in terms of p, we notice that it is a cubic equation and will have three roots. The first two roots can easily be determined by the part of the equation outside of the brackets. If p is 0, then the whole expression is 0. If p is 1, then $(1-p)$ is 0 and the whole expression is 0. The third root is determined by setting the bracketed terms equal to 0 and solving for p.

$$0 = W_{ab} - W_{bb} + p(W_{aa} + W_{bb} - 2W_{ab})$$
$$p(2W_{ab} - W_{aa} - W_{bb}) = W_{ab} - W_{bb}$$
$$p = \frac{W_{ab} - W_{bb}}{2W_{ab} - W_{aa} - W_{bb}}$$

Our three roots of p for which our population is at evolutionary equilibrium are

$$p^* = 1$$
$$p^* = 0$$
$$p^* = \frac{W_{ab} - W_{bb}}{2W_{ab} - W_{aa} - W_{bb}}$$

Note that the superscript * implies an equilibrium value. To go from no answer to three answers is a bit overwhelming. Let us look at our results to see if they have any biological meaning. The first two results $p^* = 0$ and $p^* = 1$ are known as *trivial roots*. This disparaging appellation may stem from the mathematical derivation, but not from the biological interpretation. There is a simple but extremely important concept that we derive from these answers. When p is 0 or 1, our population is made up of all bb or aa individuals, respectively. Under these conditions of genetic homogeneity, there will be no evolution caused by selection. This point may be clearer if you remember that selection is driven by relative fitnesses. Except for a lethal genotype, where fitness is zero, any genotype has a relative fitness of one when it is the only genotype. A genotype in a population cannot be selected against, relative to

Lecture 3   Selection in Natural Populations

another genotype which does not exist in the population, no matter how elegantly adaptive that nonexistent genotype might be. This means that, even if, for example, the *bb* genotype has a higher fitness than either the *aa* or *ab* genotypes, a population composed of all *aa* individuals will not evolve to be composed of all *bb* individuals. It will remain where it is. *Selection only acts on existing genetic variation; it does not create adaptations.*

The last solution for p^*, $(W_{ab} - W_{bb})/(2W_{ab} - W_{aa} - W_{bb})$, is called the *nontrivial root*, a name with which you may very well agree. It is obvious that its actual value will vary from population to population and environment to environment depending on the actual fitness constants. To complicate matters further, this nontrivial value of p^* may actually be outside the range of possible values of p, depending on the fitness constants. Frequencies must be within the range from 0 to 1, yet without too much trouble, we can all make up values for W_{aa}, W_{ab}, and W_{bb} such that p^* will be negative or greater than one. Therefore, not only will the nontrivial equilibrium value be different for each case, it may not even exist within the biologically meaningful range.

The distance that we have covered is actually farther than you may have realized. All our equilibrium values of gene frequencies describe populations that are not evolving or have stopped evolving. All other values of gene frequencies will give some positive or negative value of Δp, that is, all other gene frequencies will change under constant selection over time until they reach one of our p^* frequencies. Knowing how to determine p^* values, we can predict the outcome of selection regimes. In Lecture 4 we will discuss how to decide which of the three equilibrium values will actually be realized.

Let us review the essential points of this lecture. If different genotypes have different contributions to the establishment of the next generation, we say that they have differences in *fitness* and that *selection* is acting on the population. The fitness of an individual is a combination of three general components: *viability, mating success*, and *fecundity*. When we introduce the *coefficients of fitness* into our basic model, we understand that the magnitude of change in frequency of a given trait is determined both by the present frequencies in the population and the strength of selection. Last, a population under selection will reach an evolutionary equilibrium, that is, will stop evolving whenever the gene frequency is equal to zero or one, when there is no variation to select on, or when its gene frequency is equal to the nontrivial equilibrium point

$$\frac{W_{ab} - W_{bb}}{2W_{ab} - W_{aa} - W_{bb}}$$

lecture 4
Fisher fundamental theorem

At this point we may feel a bit like Penelope with several suitable equilibrium frequency values from which to choose. Like Penelope, you may feel that you do not want to choose. In this lecture we will discuss the criteria for making our choice: how selection works in populations, and four basic cases that occur under constant selection.

Let us exercise our intuition again. What will be the result of selection on the population as a whole as we follow it through time? While we must remember that selection is acting on individual organisms, it is not too difficult to extrapolate to the entire population. Differences in fitness by definition mean that we will see differential representation of particular genotypes in the following generation. Therefore, more individuals in the population will have the higher fitness values in each successive generation. The *average fitness* $\bar{W}$ of a population should increase. But should we expect this to continue *ad infinitum*? As discussed in the last lecture, selection does not create new adaptations, but only uses whatever already exists. Thus, we should expect that the average fitness will stop increasing when it reaches the maximum fitness possible in the population. We now have to work toward two intuitive predictions. Let us see how close we get.

Our first step in testing our intuition and in helping to solve our dilemma of too many equilibrium frequency values is to derive what is commonly known as the *Fundamental Theorem of natural selection*. This theorem, originally derived by R. A. Fisher in 1930, describes how we can expect the average fitness of a population to change over time under constant selection pressures,

Lecture 4 Fisher Fundamental Theorem

the very question that we intuitively addressed earlier. We begin by defining the average fitness of a population as we did in the last lecture.

$$\bar{W} = p^2 W_{aa} + 2pq W_{ab} + q^2 W_{bb}$$

To find out how $\bar{W}$ changes over time, we could substitute in our p' values, get $\bar{W}'$, the average fitness in the next generation, and subtract the previous average fitness:

$$\Delta \bar{W} = \bar{W}' - \bar{W}$$

We know, however, that in each generation Δp may change depending on the present value of p. So each time we must substitute new values of p' and repeat the calculation. This is a tedious process. One may easily write a computer program to implement this process and generate the evolutionary trajectory, or path, with any fitnesses and initial frequencies. But we are not always interested in the trajectory. What if we only want to know the general direction of the process and the end result? We will see shortly that we can answer these questions analytically without the help of computers.

An easier way to attack this problem is to describe the process of change as a continuous process, where each change is small and the process is smooth. In physics we do this by collapsing our time frame. For example, if we want to describe the process by which velocity changes per unit time, Δv, we make our unit of time smaller and smaller until the change in time Δt approaches zero. When we do this, Δv, our discrete change in velocity per unit time, approaches the derivative of velocity by time, or

$$\frac{\Delta v}{\Delta t} \longrightarrow \frac{dv}{dt}$$

In this case each Δv is small and the process can be seen as uniform and smooth. In biology we cannot reduce our unit of time to let it approach zero, because our time frame is fixed. Our unit of time must be one generation. We can get around this problem and cause the change in average fitness over time to be a smooth process by assuming that our changes in allelic frequencies are very small. Remember that this will happen when the differences in fitnesses of the genotypes are very small. In other words, under weak selection,

$$\frac{\Delta p}{\text{Generation}} \approx \frac{dp}{dt}$$

Most differences in fitness of genotypes in natural populations are small, so this assumption may not be too far from the truth. This, by the way, is why measuring relative fitnesses in nature is so difficult.

We can now attack our problem by looking at it as a continuous process where

$$\Delta \bar{W} \approx \frac{d\bar{W}}{dt}$$

Since $\bar{W}(t) \equiv \bar{W}(p(t), q(t))$, we can expand this by the chain rule

$$\frac{d\bar{W}}{dt} = \frac{\partial \bar{W}}{\partial p} \cdot \frac{dp}{dt} + \frac{\partial \bar{W}}{\partial q} \cdot \frac{dq}{dt}$$

$$\approx \frac{\partial \bar{W}}{\partial p}\Delta p + \frac{\partial \bar{W}}{\partial q}\Delta q$$

Taking the partial derivative, we get

$$\frac{\partial \bar{W}}{\partial p} = W_{aa}\frac{d}{dp}p^2 + W_{ab}\frac{d}{dp}2pq + W_{bb}\frac{d}{dp}q^2 = 2(pW_{aa} + qW_{ab})$$

We can simplify our notation further by remembering the concept of the marginal fitness of an allele in a population, W_a, where

$$W_a = pW_{aa} + qW_{ab}$$

For review, the marginal fitness is the average effect on fitness of a particular allele, a, in a population. We can now write our partial differentiation as

$$\frac{\partial \bar{W}}{\partial p} = 2W_a$$

Similarly,

$$\frac{\partial \bar{W}}{\partial q} = 2W_b$$

We worked out the values for Δp and Δq in the last lecture. We will use the equations for changes in frequencies which use the summary notation of marginal fitnesses.

$$\Delta p = \frac{p}{\bar{W}}(W_a - \bar{W})$$

$$\Delta q = \frac{q}{\bar{W}}(W_b - \bar{W})$$

Now we can write the change in fitness as

$$\Delta \bar{W} = (2W_a)\left(\frac{p}{\bar{W}}\right)(W_a - \bar{W}) + (2W_b)\left(\frac{q}{\bar{W}}\right)(W_b - \bar{W})$$

$$= \frac{2}{\bar{W}}[pW_a(W_a - \bar{W}) + qW_b(W_b - \bar{W})]$$

We have shown in the last lecture that the increase in frequency of one allele must be equal to the decrease in frequency of the other allele, or

$$\Delta p + \Delta q = 0$$

This is equivalent to

$$p(W_a - \bar{W}) + q(W_b - \bar{W}) = 0$$

I will now multiply this zero term by two and subtract it from the $\Delta \bar{W}$ equation. (Subtracting zero from a quantity obviously does not change it.)

$$\Delta \bar{W} = \frac{2}{\bar{W}}[pW_a(W_a - \bar{W}) + qW_b(W_b - \bar{W})]$$
$$- 2[p(W_a - \bar{W}) + q(W_b - \bar{W})]$$

We now rearrange our equation in the next few steps:

$$\Delta \bar{W} = \frac{2}{\bar{W}}[pW_a(W_a - \bar{W}) + qW_b(W_b - \bar{W}) - p\bar{W}(W_a - \bar{W}) - q\bar{W}(W_b - \bar{W})]$$

$$= \frac{2}{\bar{W}}[p(W_a - \bar{W})(W_a - \bar{W}) + q(W_b - \bar{W})(W_b - \bar{W})]$$

$$= \frac{2}{\bar{W}}[p(W_a - \bar{W})^2 + q(W_b - \bar{W})^2]$$

Some of you will recognize that the bracketed term is the *mean squared deviation* of allelic fitnesses from the average fitness, or the *variance* of allelic fitnesses in a population. Variance in statistics is denoted by the symbol σ^2. We may denote that our variance is a variance of fitness by adding a subscript W. Therefore, we may sum up the Fundamental Theorem of natural selection as

$$\Delta \bar{W} = \frac{2}{\bar{W}} \sigma_W^2$$

or the change in fitness is directly proportional to the variance of fitness in a population.

What can we conclude from this last equation? Variance can be viewed as a measure of how different individuals are in a population. In our case this difference is in terms of fitness and the individuals are organisms in a population. If the individuals are very similar, the variance will be low and the expected change in average fitness will be low. What if there are a few individuals with vastly different fitnesses from all others in the population? The variance will still be low, because, on the average, any two individuals will be very similar, and the expected change in fitness should again be small after one generation. These arguments should sound familiar. They summarize our interpretation of the expected change in frequency that we discussed in the last lecture. If p is close to 0 or 1, ΔW is small no matter what the fitnesses of genotypes.

The second interpretation which we can make from the Fundamental Theorem of natural selection is that the change of average fitness of a population under constant selection will always be positive or zero. We can see this by noticing that all terms are greater or equal to zero.

Let us compile everything which we developed from these equations and apply it to the problem of predicting the equilibrium composition of a population under selection. First, we know from the last lecture that if there is a differential in fitnesses of alleles, selection will cause a change in allele frequencies each generation. Second, it is obvious that if we change the composition of a population, we will also change the average fitness of that population. We know that, third, the change in average fitness must always be positive or zero, so that any change will increase the average fitness of the population. Fourth, we will continue to increase average fitness until any change in allele frequencies would cause a decrease in fitness, that is, until we reach a maximum average fitness. At this point, selection will no longer change allele frequencies or average fitness. We will come to rest at an evolutionary equilibrium.

All we need to know is at what value of p, that is, at what allelic frequencies, will average fitness be maximized? Remember from elementary calculus that we can determine what value will maximize a function by taking the first derivative of that function, setting it to zero, and solving the equation. The solution will correspond to a maximum or a minimum value for the function, in this case, average fitness. We must then check the second derivative with this value. If the second derivative is negative, we have a maximum value; if it is positive, we have a minimum value. Let us actually take the derivatives of the fitness function and see what is the maximum value.

We start from our original equation of average fitness in terms of one variable:

$$\bar{W} = p^2 W_{aa} + 2(p)(1-p)W_{ab} + (1-p)^2 W_{bb}$$

Taking the first derivative in terms of p, we get

$$\frac{d\bar{W}}{dp} = 2pW_{aa} + 2(1-p)W_{ab} - 2pW_{ab} - 2(1-p)W_{bb}$$

Collecting terms, we simplify this to

$$\frac{d\bar{W}}{dp} = 2[(W_{ab} - W_{bb}) + p(W_{aa} + W_{bb} - 2W_{ab})]$$

We set this equal to zero and solve for $\hat{p}$, our maximum/minimum fitness allelic frequency.

$$0 = 2[(W_{ab} - W_{bb}) + p(W_{aa} + W_{bb} - 2W_{ab})]$$

$$p(2W_{ab} - W_{aa} - W_{bb}) = (W_{ab} - W_{bb})$$

$$\hat{p} = \frac{W_{ab} - W_{bb}}{2W_{ab} - W_{aa} - W_{bb}}$$

We have now made a suitably satisfying discovery. Our maximum/minimum frequency value is equal to our nontrivial equilibrium value. This gives further confirmation of our interpretation of the Fundamental Theorem. At a fitness maximum (or minimum), we see that p, the allelic frequency, will not change.

Will selection always drive our population to an equilibrium frequency of $p^* = (W_{ab} - W_{bb})/(2W_{ab} - W_{aa} - W_{bb})$? No, we cannot say this. First,

$$p^* = \frac{W_{ab} - W_{bb}}{2W_{ab} - W_{aa} - W_{bb}}$$

may correspond to a minimum rather than maximum of fitness. From whatever allelic frequency our population started, we would then have to decrease average fitness to reach this point. Clearly, the Fundamental Theorem will not allow this. When, indeed, is this the case? Taking the second derivative of our fitness function, we see that when

$$\frac{d^2\bar{W}}{dp^2} = 2(W_{aa} + W_{bb} - 2W_{ab})$$

is greater than zero, we will have minimum fitness at p^*. The second problem is that

$$p^* = \frac{W_{ab} - W_{bb}}{2W_{ab} - W_{aa} - W_{bb}}$$

may fall outside the range of 0 to 1 and, therefore, have no biological meaning. Mathematically, it will still be the point of maximum or minimum of the fitness curve extended beyond the 0 to 1 range.

We now have all the tools to answer our questions simply. We plot the values of $\bar{W}$ as a function of p under all possible conditions and apply our conclusions from the Fundamental Theorem for the analysis.

Although the graphic analysis will be saved for the next lecture, let us at this point try to make reasonable guesses as to what equilibrium p value will give us a fitness maximum. We can arrange four possible relationships for unequal fitness coefficients.

Case I: $W_{aa} > W_{ab} > W_{bb}$
Case II: $W_{aa} < W_{ab} < W_{bb}$
Case III: $W_{aa} < W_{ab} > W_{bb}$
Case IV: $W_{aa} > W_{ab} < W_{bb}$

In Case I we can see a clear advantage for the a allele. If we compare the marginal fitnesses of the alleles, it can be shown that

$$pW_{aa} + qW_{ab} > pW_{ab} + qW_{bb}$$

or

$$W_a > W_b$$

If this is the situation, it seems intuitive that our population will reach the equilibrium frequency

$$p^* = 1$$

and will have only aa homozygotes, the fittest genotype.

Case II seems to be identical to the preceding situation except that now the b allele is the most fit. (It should be clear that it is totally arbitrary which allele we call a and which we call b.) Following the same logic, our equilibrium frequency of a should be

$$p^* = 0$$

or all our population will be bb homozygotes.

Case III is somewhat different. Now the heterozygote is the fittest genotype. Our intuition might suggest that we will maintain the heterozygotes. If this is true, then we must have both *a* and *b* alleles in our population.

It could be that

$$p^* = \frac{W_{ab} - W_{bb}}{2W_{ab} - W_{aa} - W_{bb}}$$

is intermediate between zero and one and we will maintain both alleles. What about the distribution of genotypes? We cannot maintain all heterozygotes under random mating. Think about this now, and later in the next lecture we will discuss it further.

What about Case IV? How can we select for both homozygotes? Possibly we can maintain both in the population, and our p^* will be

$$p^* = \frac{W_{ab} - W_{bb}}{2W_{ab} - W_{aa} - W_{bb}}$$

as in Case III. Then again, we will have less fit individuals being maintained by panmixia as in Case III. Perhaps the population will be driven to either trivial p^* values, zero or one, so that the population will be made up of only *aa* or *bb* homozygotes, depending on which is fitter. This is intuitively sensible. Unfortunately, it is also incorrect.

At this point, I will close as any good Victorian novelist would close his or her chapter. We have resolved some of the questions that we posed at the end of the previous lecture, but, in so doing, we have now established new problems. We have learned through Fisher's Fundamental Theorem that selection-driven evolution will have the largest changes in average population fitness when the variance of fitness is the greatest. The Fundamental Theorem also states that average fitness cannot be reduced through constant selection so that the population should reach evolutionary equilibrium at a fitness maximum. However, we see that less fit individuals may be maintained in the population, and perhaps the most fit genotypes may be selected out of the population. This quandary will be resolved in Lecture 5.

lecture 5

How selection works

In the last lecture we set the stage for our final analysis of the effect of constant selection on natural populations. We summarized the possible ways in which fitnesses may differ from one another by the following inequalities:

Case I: $W_{aa} > W_{ab} > W_{bb}$
Case II: $W_{aa} < W_{ab} < W_{bb}$
Case III: $W_{aa} < W_{ab} > W_{bb}$
Case IV: $W_{aa} > W_{ab} < W_{bb}$

Finally, we discussed what our intuition predicts and the possible problems that may arise. Let us proceed and check our predictions.

We will begin with Case I. When we compare fitnesses of the genotypes, the homozygote aa is superior in fitness to the homozygote bb and the heterozygote is of intermediate fitness between the two. We may summarize this in the inequalities $W_{aa} > W_{ab} > W_{bb}$. If we solve for the equilibrium frequencies of alleles a and b, we get

$$p^* = \frac{W_{ab} - W_{bb}}{(W_{ab} - W_{aa}) + (W_{ab} - W_{bb})}$$

$$q^* = \frac{W_{ab} - W_{aa}}{(W_{ab} - W_{aa}) + (W_{ab} - W_{bb})}$$

With a little thought, you will see that the numerator of p^* is positive, the numerator of q^* is negative, and the denominators of both are identical but

29

of unknown sign. It is obvious that either p^* is negative (q^* is greater than 1) or q^* is negative (and p^* is greater than 1) since $p^* + q^* = 1$. In either case

$$p^* = \frac{W_{ab} - W_{bb}}{(W_{ab} - W_{aa}) + (W_{ab} - W_{bb})}$$

is outside the 0 to 1 range. Next, we know that $\bar{W}$ is a quadratic function. A quadratic curve is described by a parabola, a horseshoe-shaped curve where the point of inflexion (where the two arms meet) is the maximum or minimum. If that point is outside our range, then within the 0 to 1 range $\bar{W}$ must either be monotonically increasing or decreasing as p goes from 0 to 1. We may decide which is the case by noticing that at $p = 0$, $\bar{W} = W_{bb}$ and at $p = 1$, $\bar{W} = W_{aa}$. (Why?) Since $W_{aa} > W_{bb}$, we know that the curve must be monotonically increasing as shown in Figure 5.1.

For every value of p, we know that we will increase the average fitness of the population for every positive change in p. We also know that for almost every value of p, there will be a positive change. The population will stop changing only when it reaches the maximum fitness possible and that is when $p = 1$. At this time, all the individuals in the population will be homozygous aa. We say in this case that the population is "fixed" for the a alleles.

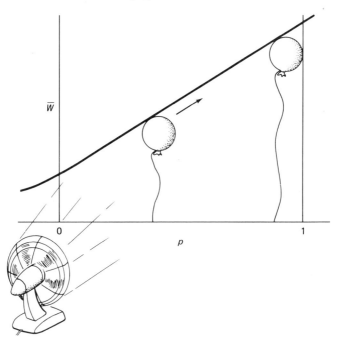

Figure 5.1 Average fitness plotted against gene frequency. In Case I where $W_{aa} > W_{ab} > W_{bb}$ the peak fitness within the range of 0 to 1 is reached at $p = 1$.

Why did I say that for *almost* every frequency of the allele a there will be an increase in frequency every generation? What frequency of a will not change? The population will not change when it is in evolutionary equilibrium and it is in equilibrium at $p = 0$, as well as at $p = 1$. When there are only b alleles, that is, when the population is fixed for b, we will not see any evolution.

At this point, it will be helpful to introduce the concept of *stability*. Let us discuss this in an informal, intuitive manner. If we can visualize our curves of average fitness as a type of ceiling between two walls at $p = 1$ and $p = 0$, and imagine our point p attached to a helium-filled balloon, we can watch how our population will evolve by releasing our balloon at a given initial frequency and following its movements. When we release the balloon, it rises to the value of average fitness at the initial frequency. However, thereafter it floats along our fitness ceiling to the highest point (being driven by the Fundamental Theorem). Looking directly below it, we will find our final equilibrium frequency. Suppose there are slight dimples in the ceiling at all equilibrium points so that the balloon will get slightly "stuck" whenever it reaches these points. If we release the balloon at $p = 0$, it will remain there. If we add to our demonstration slight air currents, the balloon may bob out of the dimples in the ceiling. The test of stability, then, is whether or not the balloon will return to the equilibrium point after being disturbed. It is obvious that if the balloon is bounced away from its $p = 1$ value to $p = 0.99$, it will float back to the original dimple in the ceiling. It is also obvious that if the balloon is bounced away from its $p = 0$ fitness dimple, it will not return but will float away toward the higher fitness. The frequency of $p = 1$ is, therefore, called a *stable equilibrium point* and the frequency of $p = 0$ is called an *unstable equilibrium point*.

The biological meaning of this analysis can be seen if we consider the air currents in real populations. If we are at some intermediate frequency of a and have a discrete number of organisms in our population, any death or birth of an individual will cause a slight change in frequency. Also, if there is migration or mutation, our frequencies can instantaneously be buffeted out of our ceiling dimples. When we analyze where our population will reach evolutionary equilibrium, we must also take into consideration which points are stable and which are unstable equilibria. The unstable equilibria will not be realized since deviations are always present.

Case II is the mirror image of Case I. Here the fitnesses of the diploid genotypes are $W_{aa} < W_{ab} < W_{bb}$. When we calculate the nontrivial equilibrium values for p and q, we again get $(W_{ab} - W_{bb})$ and $(W_{ab} - W_{aa})$ in the numerators, respectively. Again, they are of opposite signs, one positive and one negative. Their denominators are equal, so either p^* is negative and q^* is greater than 1, or q^* is negative and p^* is greater than 1. In either case these nontrivial equilibria are outside the range of frequencies (0 to 1) and are essentially nonexistent. As in Case I, the fitness curve must be monotonic between $p = 0$ and $p = 1$. When $p = 0$, $\bar{W} = W_{bb}$ and when $p = 1$, $\bar{W} = W_{aa}$. Since

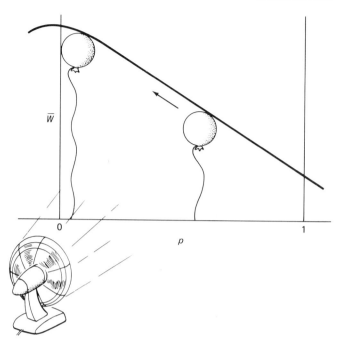

Figure 5.2 Average fitness plotted against gene frequency. Case II represents conditions where $W_{aa} < W_{ab} < W_{bb}$. The peak fitness is reached at $p = 0$.

$W_{bb} > W_{aa}$, the fitness curve will decrease as p, the frequency of the a allele increases (Figure 5.2).

The point $p = 1$ is our unstable equilibrium point (our balloon will float away from it) and the point $p = 0$ is the stable equilibrium point. Therefore, a population with these fitness values will usually become fixed for the b allele. Only when the population is composed of all aa homozygotes and there are no genetic perturbations, will the population remain at the unstable equilibrium $p = 1$.

Case III is more interesting and slightly unintuitive. In this situation the heterozygote has the highest fitness and the relationship between genotype fitnesses may be written $W_{aa} < W_{ab} > W_{bb}$. This condition, where the heterozygote fitness is beyond the range of the homozygous fitnesses, is called *overdominance*. The nontrivial value for p^* can again be written

$$p^* = \frac{W_{ab} - W_{bb}}{(W_{ab} - W_{bb}) + (W_{ab} - W_{aa})}$$

and the nontrivial q^* is, therefore,

$$q^* = \frac{W_{ab} - W_{aa}}{(W_{ab} - W_{bb}) + (W_{ab} - W_{aa})}$$

This time both numerators are positive. Furthermore, the denominator, made up of the sum of the individual numerators, is positive and must be greater than either numerator. Therefore, p^* and q^* are some frequencies between 0 and 1. We know also that at these frequencies the population will have either the highest or lowest average fitness possible. By checking the value of the second derivative

$$-2(2W_{ab} - W_{aa} - W_{bb})$$

we see that this point is a maximum. Our curve of average fitness as a function of p is shown in Figure 5.3.

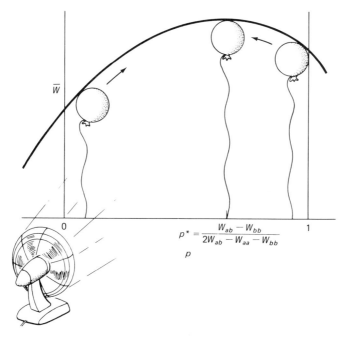

Figure 5.3 Average fitness plotted against gene frequency. In Case III the heterozygote has the highest fitness, $W_{aa} < W_{ab} > W_{bb}$. The peak fitness is reached at a nonzero or non-one frequency.

We now must contend with all three equilibrium points:

$$p^* = 0 \qquad p^* = 1 \qquad p^* = \frac{W_{ab} - W_{bb}}{(W_{ab} - W_{aa}) + (W_{ab} - W_{bb})}$$

Let us release our Fundamental Theorem balloon and add a gentle breeze. We can see that points $p^* = 0$ and $p^* = 1$ are *unstable equilibria* and the nontrivial p^* is the single stable equilibrium point (Figure 5.3). Our population will evolve to maintain a nonzero or non-one equilibrium frequency of alleles a and b.

This is more complicated than Cases I or II, but is it counterintuitive? We can review our thoughts from Lecture 4. First, our intuition tells us that the most fit genotype will completely dominate the population. This was true in Cases I and II where the populations became fixed for alleles a and b, respectively. We should expect then that our population will become fixed for the heterozygous genotype. Think for a moment what happens when two heterozygotes mate. Their offspring will have aa, bb, and ab genotypes. The less fit genotypes are constantly being reintroduced into the population. Our equilibrium population will not be fixed for one genotype, but will maintain all three genotypes. This distribution of genotypes is called a *balanced polymorphism* (*poly* = many; *morph* = types). We can also characterize the different types of selective regimes of Cases I and II as *directional selection*, leading to *monomorphic* (*mono* = one) populations and of Case III as *balanced* or *stabilizing selection* leading to *polymorphic* populations, respectively.

In Case IV the heterozygote is *underdominant*, that is, the fitness of the heterozygote is less than the fitnesses of either homozygote. We may write the relationship as $W_{aa} > W_{ab} < W_{bb}$. Note that we do not know the relationship of W_{aa} to W_{bb}. If we were to analyze the nontrivial p^* and q^* values as we did earlier, we would see that both numerators are negative. However, the denominators are the sum of both numerators and thus are also both negative. Therefore, both p^* and q^* are positive and, since $p^* + q^* = 1$, each must be less than one. We again have three potential equilibrium points to contend with in our system:

$$p^* = 0 \quad p^* = 1 \quad p^* = \frac{W_{ab} - W_{bb}}{(W_{ab} - W_{aa}) + (W_{ab} - W_{bb})}$$

As before, we may determine whether the nontrivial p^* value is a fitness maximum or minimum by evaluating the second derivative of the fitness function.

$$\frac{d^2 \bar{W}}{dp^2} = -2(2W_{ab} - W_{aa} - W_{bb}) > 0$$

The nontrivial p^* is, therefore, a fitness minimum. The Fundamental Theorem balloon will always float away from this minimum to either of the two local maxima at $p^* = 0$ or $p^* = 1$. Therefore, the nontrivial p^* must be an unstable equilibrium and $p^* = 0$ and $p^* = 1$ are stable equilibria.

How can we determine whether our final gene frequency will be $p = 0$ or $p = 1$? It should be clear from Figure 5.4 that the balloon, released at a frequency less than p^*, cannot reach $p^* = 1$ without sinking or decreasing the average fitness of the population. Similarly, it cannot move from a frequency greater than p^* to $p^* = 0$. The final equilibrium frequency of a population in Case IV is determined by the initial gene frequencies. If p_{initial} is less than the nontrivial p^*, then p^*_{final} will be zero. If p_{initial} is greater than the nontrivial p^*, the final frequency will be $p^* = 1$ (Figure 5.4).

This type of selection is called *disruptive selection*. How curious! With the same type of organism and the same environment, we can fix two populations for completely opposite alleles. Perhaps this was happening in our example in the introductory lecture where one cage of *Drosophila* became fixed for white eyes and one for red eyes. The only difference between the

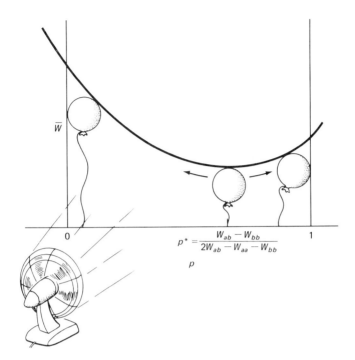

Figure 5.4 Average fitness plotted against gene frequency. In Case IV the heterozygote is less fit than either homozygote, $W_{aa} > W_{ab} < W_{bb}$. A fitness minimum is reached at a nonzero or one frequency, and a peak fitness is reached at $p = 0$ or $p = 1$, depending on initial allele frequencies.

populations in the two cages was a slight difference in initial frequency of the eye-color alleles. These initial frequencies may have been on opposite sides of a nontrivial p^* and heterozygotes may have been underdominant. Our model has indeed led to a nonintuitive result. We can characterize this result as the nonuniqueness of response. The same environment causing the same selective pressure on two populations of the same organisms with the same genotypes will produce different results.

This radically contrasts with the idea of directional evolution in which natural selection constantly tunes the population to closer adaptations to the environment. Furthermore, not only is our result nonintuitive, it is counterintuitive. Remember that we do not necessarily know which homozygote is more fit. Depending on actual values of fitness, the nontrivial p^* value will move along the p axis, but the general shape of the fitness curve will not change. Therefore, depending on the initial frequencies of the population, we can have fixation of the less fit homozygote. We have now shown how to get "survival of the less fit" under constant, stable environments. It is obvious that Darwin could not have foreseen this!

We have now answered all the questions that we posed when we started talking about natural selection in populations or organisms. We have seen that selection is a potent evolutionary force that can cause significant changes in allele frequencies in a generation. In fact, knowing nothing about actual fitness coefficients, we can fit almost any observed change to a selection model. This, of course, is its strength and its weakness. We have learned that constant selection will not cause changes in frequency of alleles indefinitely, but that populations will reach one of three evolutionary equilibria, two where there is no variation in the population and one possible polymorphic equilibrium point. The actual equilibrium point can be predicted by first determining whether or not there is variation in the population to select on, and then by determining which point represents a local maximum of fitness. To determine where such local maxima exist, we have developed a simple method of analyzing the relationships of fitness values of the genotypes and assigning them to one of four categories. In Cases I, II, and III initially variable populations will go to one determinable equilibrium point, different for each case. In Case IV we must determine the value of the unstable equilibrium point and then relate our initial allele frequencies to that point to make the prediction.

We will gain a deeper understanding of these findings in the problem-solving section that follows. A small caveat is in order first. Voltaire's Dr. Pangloss firmly believed and preached that we are in the best of all possible worlds. When faced with disease and earthquakes, he was forced to justify their existence. With the powerful tool of selection, we too may be quick to believe that all evolutionary change is the best possible; after all, selection drives populations to fitness maxima. The real world, however, is not constant and populations may be driven to fitness maxima of previous environments that are not the maxima of present environments. We can think anthropomorphically of natural selection aiming at a target only to find that the target has moved in the interval of time between aiming and firing. The environment is *always* changing, always moving the target to some extent. The population is then left to chase this moving target, sometimes closing in, sometimes falling behind, sometimes coming to underdominant forks in the road and splitting this way or that. Also, as mentioned before, selection does not create the raw material on which it acts, and it must "make do" with what exists—it must travel with the vehicle and tools it has on hand. Lastly, there are other evolutionary forces that act on populations, some weak and some strong, and their interactions may drive evolution in unexpected ways. Therefore, let us be like Candide and approach the next two chapters with open eyes.

Problem solving

These problem-solving sections review the material in the preceding lectures and add some meat to the bones of ideas that we developed. It is extremely important to remember that all our models are abstract imitations of real biological phenomena. We must always think of our answers or conclusions in biological terms to see whether or not they make sense.

We will begin by walking through a garden. In one bed I have a border of short plants with blue flowers, behind which there is a patch of marigolds and a patch of four o'clocks. The marigolds are all uniform and have yellow and brown flowers. My four o'clocks are mixed, having plants with white flowers, ones with pink flowers, and ones with red flowers. My tastes in gardens are overly fastidious, and my intentions were not to have a rag-tag patch of four o'clocks but to have a nice, red spot to contrast with the blues and yellows in the bed.

By poking around, I count 36 white, 23 pink, and 25 red flowering plants, an annoyingly small number. The flower color in these plants is determined by a one-locus, two-allele system, where red is a homozygous RR, white a homozygous rr, and pink the heterozygous Rr. What are the frequencies of the alleles R and r in the flower bed?

> Frequencies are determined by a simple division of number of occurrences of a particular trait by the total number of occurrences of all traits. If we are interested in the frequency of the R allele, we know that every red-flowered plant has two R alleles and every pink-flowered plant has one. Therefore, the number of R alleles in the population is (2 R alleles/red plant) × 25 red plants + (1 R allele/pink plant) × 23 pink plants = 73 R alleles. We know that each plant has two alleles so that the number of alleles is (2 alleles/plant) × 84 plants = 168 alleles.

The frequency of the R allele is simply $f_R = 73$ R alleles/168 alleles $= 0.43$. The frequency of the r allele may be determined in a similar way. The number of r alleles in the population is (1 r allele/pink plant) × 23 pink plants + (2 r alleles/white plant) × 36 white plants $= 95$ r alleles. The frequency of r in the population is $f_r = 95/168 = 0.57$. We see that $f_R + f_r = 0.43 + 0.57 = 1.00$, which checks out our arithmetic.

I planted the four o'clocks from seed and then thinned out the young seedlings long before they flowered. My thinning-out procedure was intended to be random with respect to flower color, because the plants had not yet flowered. Of course, I assume that there is no morphological difference in seedlings of different colored flowers which would unconsciously bias my thinning. Was the original packet of seeds in Hardy-Weinberg equilibrium?

If the original population of four o'clocks were in Hardy-Weinberg equilibrium, and my thinning-out process was random with respect to flower color, my final population should also be in Hardy-Weinberg equilibrium. Is it? The way to answer this question is to compare the actual numbers of plants of each genotype in the population with the number of plants of each genotype which we would expect, given the frequency of alleles already determined, if the population were in Hardy-Weinberg equilibrium. These expected numbers are determined by multiplying the equilibrium frequencies by the total number of plants in the population, as follows:

Expected number of white-flowered plants $= (f_r)^2$ × total number of plants
$= (0.57)^2$ × 84 plants
$= 27.29$ plants

Expected number of pink-flowered plants $= 2(f_r)(f_R)$ × total number of plants
$= 2(0.57)(0.43)$ × 84 plants
$= 41.17$ plants

Expected number of red-flowered plants $= (f_R)^2$ × total number of plants
$= (0.43)^2$ × 84 plants
$= 15.53$ plants

By a quick comparison of the expected numbers and the actual numbers, 27.29 versus 36 white plants, 41.17 versus 23 pink plants, and 15.53 versus 25 red plants, we can see that our population deviates greatly from Hardy-Weinberg equilibrium. In any finite collection of individuals we would expect some deviation from the predicted values. For example, we expected 27.29 white plants. We cannot have 0.29 plants. If we had 27 or 28 plants, would this be a big deviation? Statistics gives us tools to answer these questions, but, for the present, let us use our intuitive judgment. It looks as if our original population of seeds was not in Hardy-Weinberg equilibrium with respect to flower color.

It is clear that the seed packet had an overrepresentation of the homozygous genotypes and an underrepresentation of heterozygotes. The reason why we see this distribution is unclear. It could be due to selection against heterozygotes, *underdominance*, which we discussed in Lecture 4. It could also be due to a number of other factors which have not yet been discussed. From our comparison of observed and expected frequencies we can suggest that there may be evolutionary forces, natural or artificial, acting on our population. However, we cannot realistically speculate as to which forces these are. Let us assume that there are indeed no factors acting on the garden population, but that the deviation is due only to the manufacturer's packing process.

Let us not digress too much from our major problem, the piebald patch. It was my original intention to have red flowers, but, having seen all three colors, I have decided that any one of the three colors will do as long as the patch is uniform. Whereas I may be fastidious, I am also patient concerning my garden. I have decided on a plan of action whereby I will weed out the ones I do not want before they set seed or release pollen. Then I will take the subsequent seeds from the remaining flowers and plant them the following year. I will repeat this for as many years as it takes to get the results I want.

If all the plants produce the same number of viable seeds on the average, they will all have equal fitnesses. The proportion of plants of a particular genotype that are not weeded out will then be the fitness W of that genotype. The proportion I remove is called the *coefficient of selection*, s. The relationship between fitness and selection coefficients is then simply $W = 1 - s$. If I would like my patch to be completely white how should I plan my selection scheme?

One strategy for getting all white flowers is to treat the frequency of the allele r, f_r, as equal to q from our selection models. It should be clear that, if we want to get all white flowers, we need to get an equilibrium value $q^* = 1$, that is, $p^* = 0$. One way to get $p^* = 0$ is to create a Case II environment. This was where the bb homozygote (which we are treating as our rr white-flowered plant) had the highest fitness, the aa homozygote (our red-flowered plant RR) had the lowest fitness, and the heterozygote (our pink Rr) was intermediate. To achieve this selective regime, we simply need to clip the flowers off more red than pink plants and leave the white plants alone. For example, if I cut off the flowers of 10 of the total 25 red plants, I will be reducing the seed production by 0.40. Its coefficient of selection is $s_{RR} = 0.40$ and its fitness is $W_{RR} = 1 - s_{RR} = 0.60$. If I cut off the flowers of six of 23 pink plants, its coefficient of selection is $s_{Rr} = 0.26$ and the fitness $W_{Rr} = 1 - 0.26 = 0.74$. The fitness of my white-flowered genotypes remains at 1.0, so that $W_{RR} < W_{Rr} < W_{rr}$, and I will eventually, after many generations, get an all-white patch.

If I were in a hurry, is there a way to speed up the process? All I need to do is increase my coefficients of selection. What if I only cut off red flowers and left the white and pink flowers alone? Now the fitness relationship would be $W_{RR} < W_{Rr} = W_{rr}$. This would also lead to $p^* = 0$. Here our nontrivial equilibrium point would be

$$p^* = \frac{W_{Rr} - W_{rr}}{2W_{Rr} - W_{rr} - W_{RR}} = \frac{1-1}{2(1) - 1 - (1-s)} = \frac{0}{s} = 0$$

Our population would still go to fixation for the white flowers. Can you think of another way to get an all-white patch?

What if I am willing to put up with red flowers, but simply loathe pink? If I do not disturb the white- and red-flowered plants, their fitnesses will be $W_{rr} = 1$ and $W_{RR} = 1$. If I cut off any proportion, s, of pink flowers, the fitness is $W_{Rr} = 1 - s_{Rr}$. My fitness relationship is now $W_{RR} > W_{Rr} < W_{rr}$, or Case IV. Either the red- or white-flowered plants will become fixed. I determine which ones by calculating p^* and comparing my original frequency to it. In this example

$$p^* = \frac{W_{Rr} - W_{rr}}{2W_{Rr} - W_{rr} - W_{RR}} = \frac{(1 - s_{Rr}) - 1}{2(1 - s_{Rr}) - 1 - 1} = \frac{-s}{-2s} = 0.50$$

My initial frequency of the red allele R is 0.43. This is less than 0.50 and, therefore, my garden patch will eventually be all white. Notice that the number of flowers I cut makes no difference to the ultimate outcome. The selection coefficient s_{Rr} has to be nonzero. However, the number of generations it takes to get to equilibrium will decrease strongly with an increase in s_{Rr}.

What if I decide that a white patch is too bland, that it just will not do? I would really like the red patch for an accent. Can I use the same strategies as earlier and fix my patch for the R allele?

Basically, yes, I can. If I cut off a proportion of white flowers only or cut off a proportion of white flowers and an equal or a smaller proportion of pink flowers, we will be creating a Case I environment for my patch. I will expect that the R allele will be fixed and all individuals will have red flowers.

What about creating a Case IV environment where the pink flowers are underdominant? This will work only if p^* is less than 0.43. For example, if I clip off all the pink flowers so that $s_{Rr} = 1$ and $W_{Rr} = 0$, what is the minimum proportion of white flowers I must remove for the R allele to become fixed? As we said before, p^* must be less than our initial frequency to get fixation of R. Let us set up the inequality and solve for s_{rr}.

$$p^* = \frac{W_{Rr} - W_{rr}}{2W_{Rr} - W_{rr} - W_{RR}} < 0.43$$

Because we are not disturbing the red flowers at all, $W_{RR} = 1$. Our equation is now

$$0.43 > \frac{0 - (1 - s_{rr})}{2(0) - (1 - s_{rr}) - 1}$$

$$0.43 > \frac{s_{rr} - 1}{s_{rr} - 2}$$

$$(0.43)(s_{rr} - 2) < s_{rr} - 1$$

$$(0.43)s_{rr} < s_{rr} - 1 + 0.86$$

$$(0.43 - 1)s_{rr} < -0.14$$

$$s_{rr} > \frac{-0.14}{-0.57} = 0.25$$

Note that multiplying or dividing inequalities by a negative reverses them. Therefore, if I remove all the pink flowers and at least a quarter of the white flowers every generation, my population will eventually be all red.

Is there any way in which I can get a purely pink patch?

You should immediately see that this is an impossible task in a panmictic population. If we were to select for the heterozygous pink flowers by making them overdominant, we would maintain intermediate equilibrium frequencies for R and r. As long as their frequencies were not zero and mating was random, some RR and rr offspring would be produced. Let us take an extreme example where we remove all white and all red flowers: What would be the equilibrium phenotypic distribution? Our fitness values would be $W_{RR} = 0$, $W_{Rr} = 1$, $W_{rr} = 0$. We calculate p^* and q^* as

$$p^* = \frac{W_{Rr} - W_{rr}}{2W_{Rr} - W_{rr} - W_{RR}}$$

$$= \frac{1 - 0}{2 - 0 - 0} = 0.50$$

$$q^* = 1 - p^* = 1 - 0.50 = 0.50$$

Our phenotypic distribution would be 0.25 red, 0.50 pink, and 0.25 white. Notice that this is the maximum proportion of pink-flowered plants possible—only 50%.

In a second bed I have four o'clocks which in addition to red, pink, and white have yellow and orange flowers. The yellow-flowered plants are of genotypes *YY* and *Yr* so that the yellow dominates the white. The orange-flowered plants are of genotype *YR*. I was assured that these seeds were in Hardy-Weinberg equilibrium. I planted the seeds and got the following number of different-colored flowering plants: 28 yellow, 24 orange, 19 red, 34 pink, and 15 white. What are the allelic frequencies in this population?

We begin by assigning each allele with a letter representing its frequency: $f_R = p, f_r = q$, and $f_Y = r$. We know from the Hardy-Weinberg law that the genotypic frequencies can be derived by the following expansion: $(p + q + r)^2 = p^2 + 2pq + q^2 + 2qr + r^2 + 2rp = 1$. As with the two alleles, p^2 is the frequency of *RR*, $2pq$ is the frequency of *Rr*, and q^2 is *rr*. Our new genotypes have the following frequencies: $2qr$ is *Yr*, $2pr$ is *YR*, and r^2 is *YY*.

To solve the problems, we must now calculate the observed frequencies of genotypes by dividing the number of plants of that genotype by the total number of plants and then solving for the frequency of each allele. To clarify this procedure, all information has been compiled in one table.

Phenotype	Red	Pink	White	Orange	Yellow
Genotype	*RR*	*Rr*	*rr*	*YR*	*YY* and *Yr*
Hardy-Weinberg frequency	p^2	$2pq$	q^2	$2pr$	$r^2 + 2qr$
Number of individuals	19	34	15	24	28
Observed frequency	0.158	0.283	0.125	0.200	0.233

Since red and white flowers are distinctively homozygotic, we can solve for *p* and *q* by taking the square roots of the frequencies of the red- and white-flowered plants.

$$p = \sqrt{0.158} \approx 0.40$$
$$q = \sqrt{0.125} \approx 0.35$$

We can solve for *r* by knowing that *p*, *q*, and *r* must sum to one.

$$r = 1 - p - q$$
$$= 1 - 0.40 - 0.35 = 0.25$$

We can now check our results by comparing our expected frequencies of the pink-, orange-, and yellow-flowered plants with the actual frequencies.

Frequency of pink $= 2pq = 0.283$
Frequency of orange $= 2pr = 0.200$
Frequency of yellow $= r^2 + 2qr = 0.238$

They all correspond closely to the observed frequencies.

My using artificial selection in these problems instead of natural selection should not bother you. By cutting off the flowers of one genotype, I am effectively changing its environment. The effect of this environmental change is identical to the effect of a comparable natural change. Indeed, Darwin used the evidence of artificial selection on pigeons and domesticated animals as an argument to explain the heritable nature of selected traits in populations. The process of evolution is sensitive to the changes in fitness caused by various natural or artificial agents, but is blind to the agents themselves.

Homework exercises

Read the problems slowly and carefully before you begin solving them. Time spent in reading the problem is always well spent. Most mistakes come from misunderstanding the problem's conditions. Remember, every word has a meaning; it is not there by accident. Slow, concentrated reading and a second reading is the first rule for passing tests successfully.

You should understand the following simple equations. The Hardy-Weinberg law says that the frequencies of three genotypes in a two-allelic locus are

aa	ab	bb
p^2	$2pq$	q^2

where p is the frequency of the first allele and $q = 1 - p$ is the frequency of the second. If data are in terms of genotypic frequencies, you can check for Hardy-Weinberg equilibrium by taking the square root of the homozygote frequencies to find p and q and then comparing $2pq$ with the observed heterozygote frequency. For three alleles the law has a similar form:

aa	ab	ac	bb	bc	cc
p^2	$2pq$	$2pr$	q^2	$2qr$	r^2

Here $p + q + r = 1$ as they are frequencies of three possible alleles.

For the selection theory you should remember that

$$p^* = \frac{W_{ab} - W_{bb}}{2W_{ab} - W_{aa} - W_{bb}}$$

The value of p^* may have two very different meanings, depending on whether we are dealing with Case III ($W_{aa} < W_{ab} > W_{bb}$) or Case IV ($W_{aa} > W_{ab} < W_{bb}$). The first is the stable polymorphism; the second is the threshold value separating two qualitatively different outcomes. Do not be surprised if p^* happens to be negative or greater than one. This would mean that the calculation was unnecessary; you are dealing with directional selection (Cases I and II).

Note that the formula for p^*, which is the frequency for the a allele, has W_{bb} with a minus sign in the numerator. The formula for q^* is, of course, exactly the same except that it has W_{aa} in the numerator instead of W_{bb}. Do not confuse them—this is another typical error.

Do not memorize the dynamic equation itself. If you need it, look back at Lecture 5 and the preceding problem-solving session.

1. The peppered moth *Biston betularia* is a nocturnal creature which spends the daylight hours resting on lichen-covered tree trunks. In unpolluted areas the cryptic black-and-white color pattern of these moths camouflages them from birds. In industrial vicinities, however, the lichens are killed by airborne pollutants, and the moths resting on bare tree trunks become readily visible to predators. In these areas a dark melanic form of the moth is favored. The melanic color pattern is due to a single dominant allele. We will denote the melanic allele, D, and the recessive black-and-white allele, d. Imagine that we have a population of 142 moths in a woodlot near a large city in which 92 are DD, 43 are Dd, and seven are dd. Find
   (a) the frequencies of dominant, heterozygous, and recessive individuals
   (b) the probability of obtaining a heterozygous individual by sampling at random from this population
   (c) the total number of D alleles in the population
   (d) the frequency of D alleles (f_D) and the frequency of d alleles (f_d).

2. Calculate the expected genotype frequencies if the population of moths is in Hardy-Weinberg equilibrium. Are the expected and observed frequencies close enough to allow us to assert that the population is in Hardy-Weinberg equilibrium?

3. Assuming that there is no selection acting on the population, what are the expected allelic and genotypic frequencies after two generations of random mating?

4. Let us assume that selection is operating. If homozygous melanic moths produce five viable offspring per generation, heterozygote melanics produce four offspring, and homozygous black-and-white moths produce one, what are the absolute and relative fitnesses of the genotypes?

5. Cystic fibrosis is an inherited disease in humans caused by homozygosity for a recessive allele. The disease affects about one in 1700 newborn Caucasians.

What is q, the frequency of the recessive allele? What is the frequency of heterozygous carriers of the disease?

6. Populations of several species of grasses which grow in the vicinity of copper and lead mines have developed resistance to toxic heavy metal contamination of the soil. Assume, for simplicity, that only a single locus is involved in resistance. Suppose that p, the frequency of the allele conferring resistance (R), is 0.60 for a population of bent grass, *Agrostis tenuis*, at a certain abandoned mine site. If the fitness of the resistance homozygote (W_{RR}) is equal to 1.0, the partially resistant heterozygote fitness (W_{Rr}) is 0.80, and the nonresistant homozygote fitness (W_{rr}) is 0.05, what is the present average fitness of the population?

7. Assume random mating and no migration in our example of metal tolerance in grasses. What is Δp after one generation? What is Δp if our initial frequency of the R allele is $p_0 = 0.90$? Is it greater or less than Δp with $p_0 = 0.60$? What if selection is not so intense against nonresistant genotypes, $W_{rr} = 0.20$, for example? Is Δp greater or less than the previous case (let $p_0 = 0.60$)?

8. Sickle-cell anemia is a disease caused by an amino acid substitution in the β chain of hemoglobin molecules, resulting in a sickle-shaped deformity of the red blood cells. The allele $Hb\beta^s(a)$, which is responsible for the disease when in the homozygous condition, confers resistance to malaria when in the heterozygous condition with the normal allele $Hb\beta^+(b)$. Tragically, aa individuals almost never survive past adolescence. In regions of the world where malaria is common the heterozygote has a distinct advantage over either of the homozygotes. Suppose that in a high malaria region we find the following fitness values: $W_{aa} = 0$, $W_{ab} = 1$, and $W_{bb} = 0.85$. Calculate p^*, the nontrivial equilibrium frequency of the $Hb\beta^s(a)$ allele. What is the frequency of sickle-cell anemia in this population at equilibrium?

9. Suppose that we are in a region of the world where malaria is virtually nonexistent. In this case the fitness of heterozygotes may actually be less than the fitness of homozygotes for the normal allele, since some of the offspring of two heterozygotes will be afflicted with sickle-cell anemia. Suppose that $W_{aa} = 0$, $W_{ab} = 0.90$, and $W_{bb} = 1$. What are the selection coefficients for the two genotypes s_{aa} and s_{ab}? What is p^* for this population?

10. In a population of mosquitoes a chromosomal abnormality is associated with the heterozygous condition at a certain gene locus. Individuals homozygous for either of the two alleles (call them a and b) leave more offspring than heterozygotes. Thus $W_{aa} > W_{ab} < W_{bb}$. If $W_{aa} = 1$, $W_{ab} = 0.1$, and $W_{bb} = 0.8$, what is the nontrivial p^* (the equilibrium frequency of the a allele)? Is it stable? What is the minimum initial frequency of the b alleles needed for eventual fixation of the less fit genotype?

11. In fruit flies (*Drosophila*) white eye (rr) is recessive to red eye (RR and Rr). In population cages of white-eyed and red-eyed fruit flies we note that, when the population contains high initial frequencies of white-eyed flies, the white-eyed trait is eventually fixed. At lower initial frequencies, however, the red-eyed trait becomes fixed. If $W_{rr} = 0.4$, $W_{Rr} = 0.2$, and $W_{RR} = 1.0$, consider the following:
    (a) What will be the eventual outcome of selection in a population with initial red-eyed gene frequency of $p_R = 0.5$?
    (b) What is the minimum frequency of the white-eyed allele needed for the white-eyed trait to become fixed in a population?

chapter II

Other Evolutionary Forces

lecture 6
Mutation

In the next few lectures we will investigate the ways in which other evolutionary forces, besides natural selection, can affect evolution. The word "other" does not imply that these forces are less important or that they are necessarily weaker. It also does not imply that *all* other forces are included in this chapter. Chapter III, for instance, deals exclusively with stochastic evolutionary forces.

As a starting point for each discussion, we will use the basic model that we developed in Lecture 2. We made the following assumptions concerning our model diploid population:

1. The population is large.
2. There is no mutation.
3. There is no natural selection.
4. There is no migration.
5. There is panmixia.

We asked (1) what will be the frequency of the a and b alleles at equilibrium and (2) how will the diploid genotypes be distributed after one generation? In Lectures 3 through 5 we relaxed assumption 3 and investigated the effects of natural selection on populations. In this chapter we will relax assumptions 2, 4, and 5 and see what evolutionary forces will arise. Specifically, these will be mutation, migration, and nonrandom mating.

Mutation is the process whereby the genetic material in an organism changes spontaneously. Why is this so important? All our models so far have dealt with the changes in gene and genotype frequencies. All have started out with at least two alleles. However, the result of most of the models has been to reduce the number of alleles, usually to one per gene locus per population. Yet we know that environments do not stay constant over time, and, therefore, the fitnesses of alleles change over time. What does this portend to a population that has already become fixed for one particular allele? We know that it has reached a final evolutionary equilibrium.

Let us follow this train of thought. We have said that the fitness of an allele can conceivably decrease with a change in the environment. We have been speaking of fitness in terms of relative fitness. *Absolute fitness* can also decrease. Ecologically, this is of utmost importance. Consider that if the absolute fitness decreases drastically, the population may decrease in size to a point where it cannot maintain itself. It may go extinct, the ultimate evolutionary dead end.

Mutation is the ultimate evolutionary escape route. It is the supplier of new raw material for evolution. But mechanistically, what is it? To discuss mutation in any detail, we must first discuss the composition of the genetic material of an organism. Until this point, we have assiduously sidestepped this problem by dealing with genes and alleles as boxes, not necessarily all black, but hermetically sealed, nonetheless. When speaking of alleles, we have been imagining a type of blueprint. This blueprint can be duplicated and passed on to the next generation. In the individual organism this blueprint is used to construct a gene product, which in turn builds up the appearance and functioning, or the *phenotype*, of the organism. Different variations in the basic blueprint are our different alleles. The ultimate changes in the phenotype are caused by the different blueprints. The fitness relationships of our alleles are the final assessment of how well our different phenotypes work. Pursuing this metaphor, then, a mutation is the result of a misplaced line or a miscopied number in the blueprint when it is being duplicated. The number or line is redrawn at random. As you would expect, the gene product constructed from this randomly mutated blueprint usually does not work very well, and the completed organism may fall apart or at least be severely handicapped.

The molecular basis of the genetic blueprint is well understood. The instructions for the gene products come in the form of a linear code made up of four letters. The letters of the code are four *nucleotides*, all identical except for different nitrogen bases. These are adenine, guanine, cytosine, and thymine. These nucleotides combine to form linear, polymeric molecules of DNA. The actual order of the bases forms the code. Segments of the code are transcribed into molecules of RNA, and these in turn are translated, word by word, into a sequence of amino acids. The amino acids are bonded in the order dictated by the transcribed RNA, and a polypeptide is produced. A *gene* is the string of words, or the sentence, which provides the blueprint for the formation of the polypeptide.

The key then is the coded words on the DNA molecule, which are actually a series of three bases. Herein lies the target of mutation. As when a letter of a word is changed the whole sentence may become unintelligible, so too one change in a base may alter the final gene product. Think of the possible kinds of "typos" that may exist! We can exchange one letter for another, causing a possible change in the meaning of the word, or in the string of words that make up a sentence. Such mistakes are called *base pair substitutions*. If the change does not alter the meaning of the word (that is, if the same amino acid is coded for), or if the changed meaning does not alter the sentence or gene product (that is, a functionally equivalent amino acid is coded for), the mutation is *neutral*. If the substitution codes for a functionally nonequivalent amino acid so that the gene product is changed in some significant way, the mutation is a *missense* mutation. Finally, if the change in base pair causes the gene sequence to be misread by creating or destroying stop codes, the mutation is a *nonsense* mutation.

There are also more radical mutations which affect the structure of the linear code of a gene. Imagine the three words, "HOT MAY DAY." Each word symbolizes a three-letter genetic word. If we delete one letter, the "T" for example, and maintain three-letter words, our words are now "HOM AYD AY," which does not make much sense. Similarly, if one nucleotide base is deleted or added, the entire gene will be altered. Such mutations are called *frameshift* mutations. The number of bases involved may be more than one, but the result is usually the same: significant alteration of the gene. Likewise, there is the recently confirmed phenomenon of transposable genetic elements, which are linear molecules of DNA that insert themselves into, and extricate themselves from, the long polymeric DNA molecule of the chromosomes. These transposable genetic elements, or *transposons*, can cause spontaneous mutations if they insert themselves within a segment of a gene that is usually transcribed.

This digression into molecular genetics is not meant to be an exhaustive discussion of the molecular basis of genes or of gene mutation; rather, it is hoped this will refresh your memory before we begin our discussion of mutation on the population level.

Our population genetics discussion of mutations begins with the following scenario. We have a gene made up of a long molecule of DNA. There are only two possible forms or alleles of this gene and these are differentiated by a single base pair substitution at one site. There is no functional difference between the two alleles, that is, the mutations that change one to the other are neutral mutations. We can simplify this further by saying that all other mutations change the gene products so radically that the cells or organisms which contain such mutations never develop and die before they can be considered part of the population.

For simplicity, we will use the same nomenclature that we used in the previous lecture. Our two alleles are a and b. The frequency of the a allele is p and the frequency of the b allele is q. As before,

$$p + q = 1$$

What will be the changes in frequency of p and q given that a can mutate to b and b can mutate to a? What is the evolutionary equilibrium frequency of alleles a and b given constant rates of mutation?

We begin by assigning rates of mutation of alleles a to b and b to a as μ and ν, respectively. This can be represented in terms of a reaction equation as

$$a \underset{\mu}{\overset{\nu}{\rightleftharpoons}} b$$

The actual change in frequency will be dependent not only on the rate of mutation but on the frequency of the a and b alleles as well. This is directly analogous to a chemical reaction where both concentrations of reagents and the equilibrium constants are considered when determining the direction in which the reaction will proceed. It should be clear that we are losing a alleles per generation at a rate $-\mu p$. At the same time, we are regaining them by back mutation from the b alleles at a rate of $+\nu q$. The change in frequency per generation is then the sum of the loss and gain.

$$\Delta p = -\mu p + \nu q$$

or

$$\Delta p = -\mu p + \nu(1 - p)$$

As in our selection models, we know that

$$\Delta q = -\Delta p$$

or

$$\Delta q = \mu p - \nu(1 - p)$$

This answers our first question concerning the changes in gene frequency with mutatuion. The second question, determining the evolutionary equilibrium under forward and back neutral mutation, can be answered by a straightforward manipulation of this first equation. The evolutionary equilibrium frequency is the frequency at which there will be no further changes. As in Chapter I, this can be written

$$\Delta p = 0$$

Lecture 6   Mutation

If we take our change of frequency equation and solve it for zero, we get

$$\Delta p = 0 = -\mu p^* + v(1 - p^*)$$
$$= v - p^*(v + \mu)$$
$$p^*(v + \mu) = v$$
$$p^* = \frac{v}{v + \mu}$$

As opposed to our equilibrium frequencies under selection, we have only one root to the equation. Furthermore, we have not called on any maximizing principles as we did under selection. The reason for this has been stated before but bears repetition: Mutation does not drive a population in any predetermined direction or toward any adaptive state.

You may argue that while our p^* does not necessarily maximize anything, mutation does drive to an equilibrium. This is true, but it is true only because of the assumptions we made. We stated that we would only be concerned with two alleles. Mutation is a random process, however, acting at individual base pairs in a gene which may be several thousand base pairs long. A better set of assumptions is that we have a particular allele a and a whole class of alleles $\{b\}$, which is made up of all alleles other than a. We can still use the reaction equation as before, but now v and μ will have greatly different values. Whereas before μ was the rate at which one particular nucleotide base changes to a different particular base, now it is the rate at which the allele a undergoes any type of mutation. This rate μ will be significantly larger than under our previous assumptions. In counterdistinction, v will be the specific change in taking back a sequence to the original allelic state. This change is a rare event, since it depends on particular mutation events to occur. Thus the back mutation rate v will be very close to zero, and we may rewrite our equilibrium frequency as

$$p^* = \frac{v}{v + \mu}$$
$$\approx \frac{0}{0 + \mu}$$
$$\approx 0$$

Mutation alone will not maintain a particular allele in a population.

We have already come to two important conclusions concerning mutation in populations. First, mutations do not lead to a directed evolutionary result, and second, mutation alone will not maintain an allele in a population. We now ask how strong is mutation in changing gene frequencies as compared to selection? If both forces are pushing an allelic frequency in the same direction, we cannot determine their relative effects. A better experiment would be if they are acting in opposite directions. For example, suppose I am studying the structural product of a particular locus. Because of severe molecular constraints, all mutations to allelic forms different from the common form are lethal when found in a homozygous condition. There are, however, mutations occurring at all possible sites at a constant, given rate. What will be the frequency of the class of all mutant alleles?

Before going to the direct analytical answer, let us exercise our intuition. The situation is exactly like the previous discussion of mutation rates. We have one allele, which we will call a, and a whole class of alleles which are distinguished by being non-a. Collectively, we will call these alleles b. If μ is the rate of mutation from a to b and v is the rate from b to a, we can assume that μ is much larger than v and that v essentially approaches zero. The equation for the change in frequency of allele a is then

$$\Delta p = -\mu p + v(1 - p)$$
$$\approx -\mu p$$

What can we say about the order of magnitude of the change in frequency by mutation alone? The rate of mutation μ per locus has been measured experimentally and is generally considered to range from 10^{-8} to 10^{-5}. Since p must be between one and zero, we can assume that Δp must be on the order of 10^{-5} or smaller.

Selection, on the other hand, can be much stronger. In Lecture 3 we derived a simplified formula for the change in frequency of an allele a, under the condition of total dominance and selection against the homozygous recessive. This is similar to the hypothetical situation stated earlier. The equation is

$$\Delta p \approx sq^2 p$$

where s is the coefficient of selection against the recessive homozygote. Since s can be any number between zero and one, the Δp can potentially be on the order of 10^{-1}. This is 10,000 times greater than the highest order of magnitude caused by mutation alone! Granted this is the highest Δp possible with selection and s can be quite small; however, even under moderate selection where s is between 10^{-2} and 10^{-3}, selection will far outweigh mutation as an evolutionary force.

We can proceed quickly with our analytical solution to the problem. We have the two formulae for Δp under selection and mutation considered separately. They are of opposite signs, that is, mutation is introducing new alleles and selection is screening them out. To determine the equilibrium frequency of our common and mutant alleles, we solve for $\Delta p = 0$, balancing the mutation and selection effects.

$$\Delta p = -\mu p + sq^2 p = 0$$
$$q^2 s = \mu$$
$$q^2 = \mu/s$$
$$q^* = \sqrt{\mu/s}$$
$$p^* = 1 - q^* = 1 - \sqrt{\mu/s}$$

The frequency of the mutant class q^* is maintained at a frequency which is the square root of the ratio of mutation rate over selection coefficient. When the trait is lethal, as suggested in the scenario given, $s = 1$ and

$$q^* = \sqrt{\mu}$$

The equilibrium frequency of lethal alleles will be the square root of the mutation rate. When μ is between 10^{-5} and 10^8, q^* will be 10^{-3} to 10^{-4}.

It has been suggested in the past that controlled mating and sterilization can eliminate unwanted alleles from human populations. This philosophy comes under the general name of *eugenics*. Besides engendering grave ethical and moral problems, you can see that with recessive traits it will not even work. There will always be some mutation-regenerated alleles in the population. These will be harbored mainly in the heterozygote condition and will appear with a frequency that is greater than or equal to $\sqrt{\mu}$.

Reflect a bit on this last conclusion. The model that we have developed is, of course, simple and has many approximations. The results, however, do contain important information for any citizen of the modern world. By increasing the mutation rates through environmental degradation, such as mutagenic chemical and nuclear wastes, we are also increasing the frequency of deleterious alleles maintained in the population. If the traits are lethal, the carriers will be the heterozygous individuals. They will occur in the population at a frequency

$$2p^*q^* = 2(1 - \sqrt{\mu})(\sqrt{\mu})$$
$$= 2(\sqrt{\mu} - \mu)$$

Since $\sqrt{\mu}$ is much greater than μ, $2p^*q^* \approx 2\sqrt{\mu}$. If the trait is not totally lethal, this frequency is higher. You should be aware of this when reading reports on the expected increase of mutation rates in our surroundings.

Let us briefly review this lecture. Mutation provides an evolutionary escape for all populations by constantly creating new alleles on which selection or other forces may act. It is not, however, a directional force leading to any maximization. Furthermore, mutation alone cannot maintain an allele in a population, and, when compared to selection, it is a very weak force.

Lecture 7 will take up the subject of migration as an evolutionary force. You will see that selection is not the only strong force affecting the evolution of a population.

lecture 7
Migration

As I hope to convince you shortly, migration is extremely important in determining changes in gene frequencies. Yet the process of migration is so simple: It involves the movement of a copy of a gene from one population to another.

Without belaboring the point too much, let us discuss briefly the biological intricacies of migration. To have migration, we must have at least two separate populations. Migration rate is the proportion of a population that has *established itself* from another population. The biologically important and interesting word is "establish." It is not sufficient in our concept of migration for an individual to settle within the area of a different population. This would be *dispersal*. The new immigrant must first survive in its new surroundings. Second, it must mate and reproduce with its new neighbors: Its genes must become part of our panmictic population. In this strict sense migration is the incorporation of genes from one population into another, and, as such, it is often called *gene flow*.

What are some examples of migration or gene flow? Many agricultural weeds and insect pests experience high levels of gene flow. Seeds, eggs, or larvae may be transported along with the seeds of cultivars. These new propagules will be immigrants in any already existing population of these plants or animals. Animals that disperse their larvae over large distances and later settle as immobile adults experience much gene flow. Mussels are a good example of such organisms. Even human populations experience gene flow whenever marriages occur across ethnic, racial, or social boundaries (as when a Montague marries a Capulet).

Imagine the following situation. We have a large source, or mainland, from which organisms depart and a host island where they settle (Figure 7.1). We can build a model by making the following assumptions. First, to consider migration in the sense of gene flow, we must assume that new immigrants

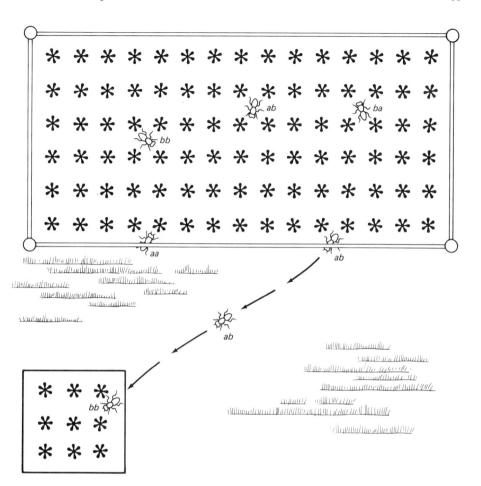

Figure 7.1 Migration from a large population to a limited host population.

become fully integrated genetically into the host population. Therefore, we can make all the assumptions that we made concerning our equilibrium or Hardy-Weinberg population, with the single exception of the one assumption concerning migration. Second, we must assume that migration is random with respect to genotype. In other words there are no genes differentially causing wanderlust among individuals. If this is true, any group of migrants from a given population is an unbiased sample of that population. We should then expect that the gene frequencies of this group of migrants will be equal to the gene frequencies of the source population. Stated more explicitly, I will call the frequency of the immigrating individuals from the population q_i. If our previous assumption is correct and q_s is the frequency in the source population,

$$q_i = q_s$$

We have all the assumptions we need, so let us proceed with our model. Any change in the gene frequencies of our island host population, q_h, from one generation to the next will be due to new immigrants. The new frequency will be the proportion of the population that are old residents times the old frequency plus the proportion of the population that are new immigrants times their gene frequency. This is a type of mean frequency. Let us call our *migration coefficient m* and let it equal the proportion of new immigrants to a population per generation. By this notation, the proportion of old residents is then $(1 - m)$. We now can put our model together. The frequency of the a allele in the host population after immigration to this population, q'_h, is

$$q'_h = (1 - m)q_h + (m)q_i$$
$$= (1 - m)q_h + (m)q_s$$
$$= q_h + m(q_s - q_h)$$

The change in frequency per generation due to migration is

$$\Delta q_h = q'_h - q_h$$
$$\Delta q_h = q_h + m(q_s - q_h) - q_h$$
$$\Delta q_h = m(q_s - q_h)$$

You can clearly see that the change in frequency is directly proportional to the coefficient of migration and to the difference in gene frequencies between the source and host populations. Will this system ever reach equilibrium in the face of constant migration? We can determine this by setting $\Delta q_h = 0$.

$$\Delta q_h = m(q_s - q_h) = 0$$
$$0 = q_s - q_h$$
$$q_h^* = q_s$$

The gene frequencies will stop changing only when the host population has the same gene frequencies as the source population. In other words in the face of constant migration a source population will eventually swamp the host or target population, no matter how small m, as long as it is not zero.

In the last lecture we introduced the idea of comparing the relative "strengths" of our evolutionary forces. Continuing with this idea, let us compare mutation and migration. The change in frequency due to mutation is

$$\Delta q = -\mu q$$

and is on the order of 10^{-6} or less. The change in frequency due to migration depends on the difference in gene frequencies between populations and the migration rate. For the sake of argument, say that the difference in gene frequencies between populations is 0.10. What rate of migration would we need exactly to counterbalance the highest expected change of frequency due to mutation? Let us set up the problem by using our formula for Δq due to migration with the values given.

Lecture 7   Migration

$$\Delta q = m(q_s - q_h) = 10^{-6}$$
$$= m(0.10) = 10^{-6}$$
$$m = 10^{-5}$$

Our migration rate necessary to balance a high mutation rate is one immigrant in 100,000. What does this value mean? If we are dealing with a population of 1000 humans, a migration rate of 10^{-5} would be equal to one immigrant entering that population every 100 generations. If a human generation is 20 years, this would be equal to one immigrant every 2000 years! I think you will agree that this is small and that actual migration rates are probably higher. Usually, we can assume that migration will swamp out the effects of mutation.

Comparing migration to selection is more difficult. As we have seen, almost any change of frequency can be attributed to selection ex post facto simply by juggling coefficients of selection. The same is true with migration: We can juggle coefficients of migration, maintaining reasonable values, and get similarly high values of Δq.

For example, there is a population of flies on an island with a gene frequency of 0.20 for a recessive allele t. This imaginary allele codes for gaudy coloration on the abdomen and on the upper half of the legs when in homozygous condition. It is known that this t, or tourist allele, is highly disadvantageous on the island, in that all the native predators differentially prey on this coloration pattern. From an enclosure study, the change in frequency of allele t due to selection, $-sp_h q_h^2$, was 0.07 per generation. However, the allele is maintained in the population by constant migration from the mainland where the frequency of the t allele is 0.80. To determine the coefficient of migration, m, we can set overall Δq due to migration and selection as

$$\Delta q_h = m(q_s - q_h) - sp_h q_h^2$$

At equilibrium, Δq is zero, so that

$$\Delta q \text{ migration} = \Delta q \text{ selection}$$
$$m(q_s - q_h) = sp_h q_h^2$$
$$m(q_s - q_h) = 0.07$$
$$m(0.80 - 0.20) = 0.07$$
$$m = \frac{0.07}{0.60}$$
$$m = 0.117$$

To balance selection, we need a coefficient of migration of approximately 12%. If our island population of flies were 100, 12 of these flies must be new immigrants for the tourist allele to persist at frequency $q_h = 0.20$. This is a rather high migration rate, but then the selective pressure was also high. Twelve percent is not an unreasonable value. You can see that migration can balance selection.

To review, migration is the incorporation of foreign genes into the gene pool of a population. The strength of its evolutionary force is directly proportional to the coefficient of migration and to the difference in gene frequencies between the source and host populations. If unopposed by other evolutionary forces, constant migration into a population will continue to change the gene frequencies of the population until it is swamped by the source population. In other words constant, unopposed migration will cause changes in gene frequencies until the two populations have an identical genetic makeup. Of the other evolutionary forces which we have discussed, mutation is too weak to counterbalance changes due to migration. Selection and migration, however, can theoretically have the same range of effects on gene frequencies and could balance each other under certain circumstances. Migration is therefore a simple but potent force in evolution.

While discussing the effects of migration on population, and ostensibly only dealing with the assumption of no migration in our Hardy-Weinberg equilibrium model, I have tampered with another assumption. For the assumption of panmixia, which states that every individual has an equal opportunity of mating with every other individual, to be true, our universe of interest must be one population. However, we must have at least two populations to be able to discuss migration. Having two populations implies that an individual in one population normally cannot mate with an individual in the other population. In other words within the present universe of interest, there is non-panmictic mating. This is only a subtle trespass on the assumption and, in the case of migration unopposed by selection, the populations become effectively one anyway. There are, however, less subtle trespasses on panmixia that occur in natural population, or within groups of natural populations. The results of these violations are fascinating and highly germane to evolutionary changes. These trespasses, lumped under the title of inbreeding, will be the subject of Lecture 8.

lecture 8
Inbreeding

The topic of inbreeding covers a multiplicity of peccadilloes, all related to nonpanmictic mating. In the narrow sense inbreeding deals with the mating of relatives above and beyond that expected by chance. In a broader sense we can discuss the results of all mating schemes that deviate from random mating patterns. These include positive and negative assortative mating, selfing and outcrossing, and mating within spatial subdivisions. More often than not, these subjects are treated separately. I would like to present them all together, since all have similar evolutionary effects and require the same conceptual framework for understanding. Once you have a good intuitive feeling for one, all the other inbreeding topics will also be clear.

The best way to conceptualize inbreeding is to think of a population at a point of time in the past which we will call our time zero. Assume that no two individuals in the population were at all related at that time. In other words there was no possibility that any two individuals share a gene received from a common parent or other ancestor. All the genes in the population can be thought of as new. All subsequent generations from this point in time will have copies of these "original" genes. After several generations, there is some probability that two gametes containing copies from the same original gene will fuse and form a diploid individual. This individual then will have two copies of the same original gene at this one particular locus (Figure 8.1). An individual that has two copies of the same gene is said to be *inbred*. The probability of getting such an individual in a population is called the *coefficient of inbreeding* and is symbolized by F. After a little consideration, you will not find this conceptually difficult. Once you have this idea, all the rest is simple.

We start again with our basic diploid Hardy-Weinberg equilibrium population and maintain all our assumptions except panmixia. Remember our equilibrium genotype frequencies:

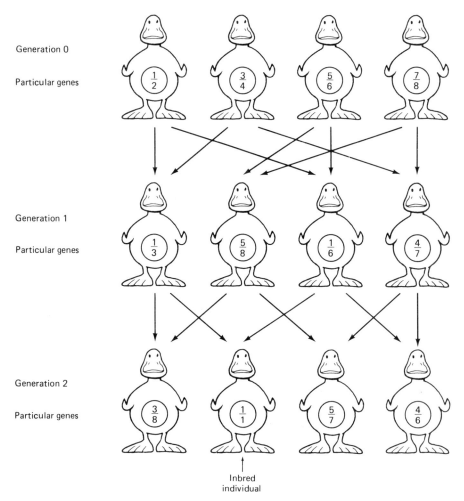

Figure 8.1 Pedigree following the fate of copies of individual genes over three generations.

$$f_{aa} = p^2$$
$$f_{ab} = 2pq$$
$$f_{bb} = q^2$$

Let us assume that there is some inbreeding in the general sense, which means that F is not equal to zero. A certain number, but not all, of the homozygous *aa* and *bb* genotypes will be homozygous not only for the same allele but also for the same gene copy. These are called *autozygous* (inbred) individuals. All the heterozygotes, of course, cannot be inbred since they obviously have copies of different genes. The proportion of the population that is not inbred is $(1 - F)$

and the proportion of the population that is inbred is F. In the noninbred part of the population the genotypes will be produced by random mating and will thus have the usual Hardy-Weinberg frequencies. Among the inbred individuals, the proportion that is (a) autozygous is equal to the proportion of the allele a in the parent population, or p, and the proportion that is (b) autozygous is equal to the proportion of allele b in the parent population, or q.

The total distribution of inbred individuals is then described by

$$pF \text{ (} a \text{ autozygous)} + qF \text{ (} b \text{ autozygous)} = (p + q)F$$
$$= F(\text{total proportion of inbred individuals in the population})$$

Thus the genotype distribution under inbreeding is

$$f_{aa} = (1 - F)p^2 + (F)p$$
$$f_{ab} = (1 - F)2pq$$
$$f_{bb} = (1 - F)q^2 + (F)q$$

This distribution covers all the possible genotypes as we can demonstrate by showing that the expected frequencies all sum to 1.

$$f_{aa} + f_{ab} + f_{bb} = (1 - F)p^2 + F(p) + (1 - F)2pq + (1 - F)q^2 + (F)q$$
$$= (1 - F)(p^2 + 2pq + q^2) + (F)(p + q)$$
$$= (1 - F)(1) + (F)(1)$$
$$= 1$$

What about the actual gene frequencies? Does inbreeding cause any change in gene frequencies from one generation to the next, that is, does inbreeding present an evolutionary force? We can test this by calculating p' from the genotype frequencies as we did in Lecture 2.

$$p' = \frac{2f_{aa} + 1f_{ab}}{2(f_{aa} + f_{ab} + f_{bb})}$$

$$p' = \frac{2[(1 - F)p^2 + (F)p] + (1 - F)2pq}{2(f_{aa} + f_{ab} + f_{bb})}$$

The sum in the denominator is equal to one, so we may simplify our equation.

$$p' = (1 - F)p^2 + (F)p + (1 - F)pq$$
$$p' = p^2 + pq + (F)(p - p^2 - pq)$$
$$p' = p(p + q) + (F)p[1 - (p + q)]$$
$$p' = p(1) + (F)p(1 - 1)$$
$$p' = p$$

Regardless of the value of F, inbreeding does not change the gene frequencies of a population from one generation to the next.

We now have a conceptual definition of inbreeding, but we still do not have a good intuitive feeling. We can develop this intuition by going back to our genotype frequencies. When coming upon a population, and having no prior knowledge of forces acting on it, we normally check to see if it is in Hardy-Weinberg equilibrium. We did this with the population of four o'clocks at the end of Chapter I. We compared the observed genotype frequencies to see if they were close to the expected Hardy-Weinberg values, given the various gene frequencies. We can do the same here. The expected frequency of heterozygotes is $2pq$. The notation

$$H_e = 2pq$$

will represent the expected proportion of heterozygotes. We previously worked out the observed frequency of heterozygotes with inbreeding. We will rewrite it with the following notation:

$$H_o = (1 - F)2pq$$
$$= (1 - F)H_e$$

Using this notation, we can derive a new definition of F as follows:

$$H_o = (1 - F)H_e$$
$$H_o = H_e - FH_e$$
$$FH_e = H_e - H_o$$
$$F = \frac{H_e - H_o}{H_e}$$

Thus the coefficient of inbreeding is but a deficiency of heterozygotes in a population relative to the expected Hardy-Weinberg values. This value of the coefficient of inbreeding is often denoted by F_{IS}; the subscript IS stands for "an individual within a subpopulation." Specifically, this coefficient of inbreeding measures the nonrandomness of mating within a population.

How can nonrandom mating reduce the number of heterozygotes in a population? Let us first look at mating between relatives, known as inbreeding in the narrow sense. Intuitively, we know that offspring from within-family matings have a higher probability of possessing two copies of the same gene than individuals within a population that are the result of random matings. This is because the two related parents contain copies of genes from common ancestors. Thus there is a chance that the fusing gametes will contain copies of the same gene. Repeated intrafamily mating increases the probability of being homozygous for the same gene copy. Similarly, the closer the relation between mating individuals, the higher the value of F.

The easiest way to conceptualize this process is to look at a system of extreme inbreeding such as self-fertilization. This is a fairly common occurrence in many plants where it is called self-pollination, selfing, or autogamy. It is

Lecture 8 Inbreeding

a less common occurrence, but by no means nonexistent, in mollusks and oligochaetes. We start by looking at a population in Hardy-Weinberg equilibrium. Again, the genotype distributions are the familiar $p^2, 2pq$, and q^2. After one generation of inbreeding without any selection, we would expect that all selfing homozygotes will produce homozygotes. However, only half the offspring of selfing heterozygotes will be heterozygous and the other half will be either of the two homozygotes. We can write the genotype distributions as follows:

$$P' = p^2 + \frac{1}{4}(2pq) = P + \frac{1}{4}Q$$

$$Q' = \frac{1}{2}(2pq) = \frac{1}{2}Q$$

$$R' = q^2 + \frac{1}{4}(2pq) = R + \frac{1}{4}Q$$

The coefficient of inbreeding can be calculated:

$$F = \frac{H_e - H_o}{H_e} = \frac{2pq - pq}{2pq}$$

$$F = \frac{pq}{2pq}$$

$$F = \frac{1}{2}$$

If we repeat this breeding system for another generation, we will again reduce the heterozygotes by half so that

$$Q'' = \frac{1}{2}Q' = \left(\frac{1}{2}\right)^2 Q$$

Thus

$$F = \frac{2pq - \left(\frac{1}{2}\right)^2 2pq}{2pq}$$

$$= \frac{2pq - \frac{1}{2}pq}{2pq}$$

$$= \frac{\frac{3}{2}pq}{2pq}$$

$$= \frac{3}{4}$$

We may generalize these equations for continuous inbreeding for n generations in terms of an original ($n = 0$) population.

$$F_n = 1 - \left(\frac{1}{2}\right)^n (1 - F_o)$$

$$Q' = (1 - F)Q$$

$$Q_n = \left(\frac{1}{2}\right)^n (1 - F_o) 2pq$$

If our original population is not inbred,

$$F_o = 0$$

$$F_n = 1 - \left(\frac{1}{2}\right)^n$$

$$Q_n = \left(\frac{1}{2}\right)^n 2pq$$

It is obvious that, as n increases, the number of heterozygotes in the population rapidly approaches zero and the coefficient of inbreeding approaches one. The frequencies of the homozygous aa and bb genotypes approach p and q, respectively. In the meantime the gene frequencies do not change.

What happens when the entire population does not self but a certain proportion of the population outcrosses? Will the heterozygous frequency go to zero? This situation occurs in many grass species where most individuals self but a few percent are wind pollinated. We will call the proportion of outcrossing in the population t. At any time n, we can determine the frequency of heterozygotes as the sum of the heterozygotes produced by random mating and the heterozygotes remaining from selfing.

$$Q_n = t(2pq) + (1 - t)\frac{1}{2}(Q_{n-1})$$

When we have reached an equilibrium, Q_n will be the same as Q_{n-1} or the genotype frequencies will not change. Remember our question: Is $Q_n = 0$? Let us solve for the equilibrium value of Q^*.

$$Q^* = t(2pq) + (1 - t)\frac{1}{2}Q^*$$

$$Q^*\left(1 - \frac{1-t}{2}\right) = (t)(2pq)$$

$$Q^*\left(\frac{2 - 1 + t}{2}\right) = (t)2pq$$

$$Q^* = \left(\frac{2}{1 + t}\right)(t)2pq$$

$$Q^* = \left(\frac{2t}{1 + t}\right)2pq$$

If t is not zero, then Q^* will always be greater than zero. Even if there is only a small percentage of outcrossing, there will still be heterozygotes in the population. In this case the relationship between the outcrossing percentage t and the coefficient of inbreeding is

$$(1 - F) = \frac{2t}{1 + t}$$

$$F = 1 - \frac{2t}{1 + t}$$

$$F = \frac{1 - t}{1 + t}$$

The purpose of these examples is to show the different ways inbreeding can occur in natural populations. What other general patterns can we expect in natural populations? Inbreeding in the narrow sense does not have to be limited to self-fertilization but, as explained previously, refers to any degree of mating between relatives. Many animals, including humans, have such a breeding system. Such mating systems are fascinating because, qualitatively, their results are the same as those given in the discussion of self-fertilization. Intuitively, you may rebel against the idea that first-cousin mating is similar to pure selfing. The difference between the two is only the speed with which F increases and the heterozygotes decrease. This is due to our original definition of the coefficient of inbreeding, the probability of being homozygous for copies of the same gene. Offspring of selfing individuals have a 50% chance of getting the same gene twice. For example, if a selfing parent is of genotype ab, where a and b are two different alleles, one-quarter of the offspring will be aa and one-quarter will be bb. Thus half the offspring are fully inbred. In parents that are less closely related than an individual is to itself this probability of fully inbred individuals is smaller.

How does this relate to our previous equations? In the pure selfing model the probability of getting two copies of the same gene is one-half and, therefore, the frequency of heterozygotes is reduced by one-half each generation. With other types of inbreeding the probability of getting two copies of the same gene is smaller. Furthermore, the actual genotypes of mating pairs are more complicated. For example, we could get a cross with $ab \times bb$ or $ab \times cd$ or $aa \times bb$ individuals. This makes the analysis more difficult. For our purposes it is not necessary to derive the general equations. A comparison of the rates of reduction given by a repeated familial mating system will suffice to drive the idea home. Table 8.1 gives the approximate reduction rate of heterozygotes for the mating patterns. They vary greatly from that of pure selfing, but all eliminate the heterozygotes, given enough time.

TABLE 8.1 Rates of Reduction per Generation of Heterozygotes in a Population

Mating	Heterozygotes as a Function of Previous Heterozygotes
Selfing	$Q' = (0.500)Q$
Full sibs	$Q' = (0.809)Q$
Double first cousins	$Q' = (0.902)Q$

We have been discussing inbreeding in terms of one locus per individual. You should understand that this is only a matter of convenience when developing our mental tools. *All* loci in an individual are affected by inbreeding in the narrow sense; all loci will tend toward homozygosity. What does this mean to a population? I have already emphasized that inbreeding alone will not change gene frequencies, but what happens when it is not alone? Remember Lecture 6 ("Mutation") when we discussed the balance between mutation and selection. The equilibrium frequency of a recessive, deleterious allele was

$$p^* = \sqrt{\mu/s}$$

Most of these alleles were maintained in the heterozygous condition and only

$$(p^*)^2 = \frac{\mu}{s}$$

individuals were phenotypically affected. If we were to add inbreeding to this system, the frequency of autozygous, affected individuals would increase dramatically. For example, it is known that, among parents of children having genetic diseases, the frequency of close within-family matings is well above average for the human population as a whole. Indeed, if you look at human populations that have much inbreeding, you will find that diseases which usually occur only rarely in the general population are much more common. Close within-family marriage is a common taboo in several cultures. These taboos are even codified into laws by many religions, as is true in Judeo-Christian traditions. Although population genetics is not necessarily the basis for these taboos, it is within the realm of speculation.

There are types of nonrandom mating, or inbreeding in our loose sense, that do not affect the entire genome. These fall under the category of *assortative matings*. These matings occur when there is an active mate selection based on some phenotypic trait. When an individual chooses a mate that is phenotypically similar, we call this *positive* assortative mating. When opposites attract, we call this *negative* assortative mating. Positive assortative mating occurs in many species of plants and animals where color, size, or behavior may be a mating stimulus. Negative assortative mating is also not rare. For example, it can be seen in matings between short- and long-styled primroses. The results of such mating patterns are parallel to the types of inbreeding discussed before. Positive assortative matings increase the homozygosity of the alleles involved and thus increase the coefficient of inbreeding F. Negative assortative mating does the opposite. It increases the number of heterozygotes above the expected number and thus decreases or gives a negative coefficient of inbreeding.

However, remember that these mating systems only affect the loci that are involved in the phenotypic expression that is stimulating or facilitating mating, or those that are tightly associated with such loci. All other loci will remain in Hardy-Weinberg equilibrium. Therefore, in a population with positive assortative mating the frequency of deleterious homozygotes would not be more common than in a randomly mating population. If the blond-haired, blue-eyed people we all see in advertisements are produced only from blond-haired, blue-eyed parents, there is no reason to think that they have a greater probability of tongue-rolling than other people.

You should now have a good, general understanding of the way mating patterns can cause deviations from expected Hardy-Weinberg genotype frequencies. Spatial subdivisions of a population can cause similar deviations. These subdivisions will produce a positive F value, that is, there will be fewer heterozygotes than we would expect, and it may look as if nonrandom mating is occurring. Restrain your intuition until we have finished our formal arguments and then we will see if all this makes sense.

Say that we are looking at a population of rabbits. The population is made up of two warrens. We look at one locus that conveniently has two alleles. We pool all our data from both warrens and get the average frequency for both alleles in the whole population. If the warrens have the same number of rabbits, we can get the average frequency of each allele, by averaging over the warrens:

$$\bar{p} = \frac{1}{2}p_1 + \frac{1}{2}p_2$$

$$\bar{q} = \frac{1}{2}q_1 + \frac{1}{2}q_2$$

We check for Hardy-Weinberg distribution and find that there is a lower frequency of heterozygotes than the expected $2\bar{p}\bar{q}$. F is positive.

Let us backtrack and see where the problem lies. If the warrens did not have exactly the same gene frequencies, then we know that there is some variance in gene frequencies. We introduced the idea of variance in Lecture 4. It is the average squared deviation from the mean.

$$\sigma^2 = \frac{1}{2}(p_1 - \bar{p})^2 + \frac{1}{2}(p_2 - \bar{p})^2$$

We can rearrange this algebraically to

$$\sigma^2 = \frac{1}{2}(p_1^2 + p_2^2) - \bar{p}^2$$

The variance of the second allele is exactly equal to this σ^2,

$$\sigma^2 = \frac{1}{2}(q_1^2 + q_2^2) - \bar{q}^2$$

Look closely at these equations. $\bar{p}^2$ and $\bar{q}^2$ are our expected frequencies of homozygotes. $1/2(p_1^2 + p_2^2)$ and $1/2(q_1^2 + q_2^2)$ are the average frequencies of homozygotes that we really see. Rearranging the equations, we have

$$\frac{1}{2}(p_1^2 + p_2^2) = \bar{p}^2 + \sigma^2$$

$$\frac{1}{2}(q_1^2 + q_2^2) = \bar{q}^2 + \sigma^2$$

We have more homozygotes in the population than we expect—one unit of variance more for each homozygote. This must mean that we have this amount less of heterozygotes, or

$$\frac{1}{2}(2p_1q_1 + 2p_2q_2) = 2\bar{p}\bar{q} - 2\sigma^2$$

If we calculate F from these values, we get

$$F = \frac{H_e - H_o}{H_e}$$

$$F = \frac{2\bar{p}\bar{q} - \frac{1}{2}(2p_1q_1 + 2p_2q_2)}{2\bar{p}\bar{q}}$$

$$F = \frac{2\bar{p}\bar{q} - (2\bar{p}\bar{q} - 2\sigma^2)}{2\bar{p}\bar{q}}$$

$$F = \frac{2\sigma^2}{2\bar{p}\bar{q}}$$

$$F = \frac{\sigma^2}{\bar{p}\bar{q}}$$

This explains the missing heterozygotes. As long as there is any variance in gene frequencies, F will be positive and there will be a deficiency of heterozygotes. This specific coefficient of inbreeding is often written F_{ST}, the subscript standing for inbreeding from "subpopulations within the total population." When we unknowingly count organisms across subpopulations or when two or more populations mix after they have mated, we always see a depression in heterozygosity. This is generally known as the *Wahlund effect*.

Try out your intuition. The structure of subpopulations allows mating only within the subpopulation. Any deviation in allele frequencies causes an increase in either p or q in any given subpopulation. This will produce a subsequent increase in either p^2 or q^2. However, the frequency of the heterozygote within a subpopulation, $2pq$, may not increase, since when p increases, q will decrease and vice versa. The easiest way to conceptualize this is to take an extreme example where there are two subpopulations. Let us say that $\bar{p} = \bar{q} = 0.5$, but $p_1 = 0, q_1 = 1$, and $p_2 = 1, q_2 = 0$. (The subscripts refer to the subpopulations.) In subpopulation 1 there are no heterozygotes and only bb homozygotes. In subpopulation 2 there are only aa homozygotes. When we sample both subpopulations we do not find any heterozygotes, which is contrary to our expectation of $2\bar{p}\bar{q} = 0.5$. The deficiency is absorbed by the homozygote classes.

Look at this same population and allow the subpopulations to intermingle, but let bb individuals mate only with other bb individuals, and aa only with other aa individuals. Effectively, we have subpopulations forming within the original population. Thus the results of the two processes are conceptually similar: There will be positive coefficients of inbreeding in both.

Lecture 8 Inbreeding

We have covered many topics in this lecture at whirlwind speed. All our discussions have centered around the probability of an individual being homozygous for copies of the same gene, that is, the probability of being inbred in the general sense. We have seen that a population can become inbred by means of nonrandom mating of any individual in a population. This results in a deficiency of heterozygotes from the expected number. The coefficient of inbreeding F in this case was

$$F_{IS} = \frac{H_e - H_o}{H_e}$$

When this nonrandom mating takes the form of matings between relatives, all loci are similarly affected. When the nonrandom mating takes the form of positive or negative assortative mating, only those loci involved in the mating decision are affected. We have also seen that population subdivisions can mimic nonrandom matings by producing a similar deficiency in heterozygotes. This time the coefficient of inbreeding can be determined by

$$F_{ST} = \frac{\sigma^2}{\bar{p}\bar{q}}$$

However, no matter what the source of the inbreeding, only the genotype frequencies are rearranged; the gene frequencies remain constant.

The last three lectures cannot be summed into a coherent one-sentence story. Clearly, mutation is the odd man out. Aside from maintaining its importance in evolution by being the ultimate provider of genetic variation, it does not rank with the rest as a potent evolutionary force. Inbreeding itself does not cause any gene frequency changes, but it is related to migration in that both processes are basically related to the organization of mating patterns. Migration deals with the effect of Ludwig marrying Shanda, while inbreeding concerns Sasha marrying Svetlana. This then leads to the question of what will happen over a long period of time when Brunhilde only marries Otto, and Maria only marries Juan. This introduces the subjects of reproductively isolated populations, genetic drift, and neutral evolution, to be discussed in Chapter III.

Problem solving

We will leave my garden (Chapter I) and go to the seashore to look at populations of barnacles, marine invertebrates. Barnacles have a pelagic larval stage during which they can be dispersed great distances by water currents. They then settle and cement themselves to hard, stable surfaces such as rock facings, pilings, or wave breakers. The particular area where our populations are found is a rocky shore interrupted by the intake and output channels of the cooling system of a power plant.

Around the area of the power plant there are two readily distinguishable morphs of the barnacle, one with solid white plates surrounding the body of the organism, and one with thin translucent plates. The two morphs differ in one enzyme in the system that deposits calcium carbonate into the plates. All genotypes can be distinguished biochemically because the ambient pH of the cells of the heterozygote is intermediate between the two homozygotes, but not all genotypes can be distinguished by the morphology of the plates.

Two populations are sampled along the coastline. One sampling site (A) is 20 miles down the coastline from the power plant and the other (B) is next to the power plant but not within the intake-output channels. Refer to the map for their locations (Figure II.1). At site A there are no thin-plated individuals and only two heterozygous individuals per 1000 sampled. At site B there are also no thin-plated individuals, but there are 15 individuals per 1000 tested that are determined to be heterozygous by their intracellular pH. This sevenfold difference in the number of heterozygotes is of interest. We should first determine the gene frequencies of the thin-plated allele, which we will call *th*, at the two sites.

The determination of gene frequencies should be a trivial problem for you by now. Let us write the frequency of the *th* allele at site A as

Chapter II   Problem Solving

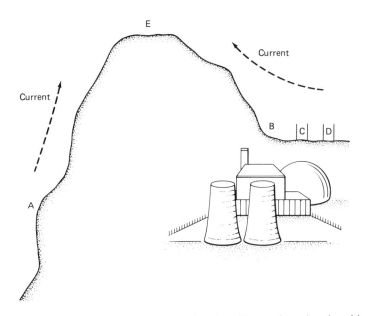

Figure II.1 Study site indicating sample locations (Roman letters) and position of power plant.

$f_{th(A)}$. In the case of no homozygotes this frequency will be equal to the number of heterozygotes divided by two times the number of individuals.

$$f_{th(A)} = \frac{2}{2000} = 0.001$$

Similarly, for site B, we can determine

$$f_{th(B)} = \frac{15}{2000} = 0.0075$$

We did not find a thin-plated individual, that is, a homozygous *th/th* genotype in either site. Is this necessarily surprising, considering our gene frequencies?

If our populations are in Hardy-Weinberg equilibrium, our expected number of thin-plated homozygotes is the square of the frequency of the *th* allele times the sample number. In our case the sample size was 1000, so

$$\begin{aligned} N_{th/th(A)} &= (f_{th(A)})^2 \times N_A \\ &= (0.001)^2 \times 10^3 \\ &= 0.001 \end{aligned}$$

$$N_{th/th(B)} = (f_{th(B)})^2 \times N_B$$
$$= (0.0075)^2 \times 10^3$$
$$= 0.056$$

In both cases the number of individuals expected is less than one. Therefore, it is not all that surprising that we did not find a homozygote in our samples.

This leads to an interesting problem. Let us assume that the populations are in evolutionary equilibrium. What is maintaining this polymorphism in the population? We learned in the previous chapter that overdominance or heterozygotic superiority can maintain a balanced polymorphism. In this chapter we learned that a mutation-selection balance, as well as a migration-selection balance, can maintain a polymorphism. To check each of these possibilities, we design a series of quick-and-dirty experiments. Let us put heterozygotes and both types of homozygotes in both sites and observe the mortality. In addition we grow them in the laboratory to get an idea of their fecundity. Last, we describe the patterns of current flow to get an idea of possible migration routes.

To begin, we must obtain the homozygous *th/th* individuals. Obviously we cannot collect enough of them from the natural populations, so breeding is necessary. Can you suggest two breeding plans that will eventually give thin-plated individuals?

> From our selection models, we know that, by artificial selection against the heterozygote, we should get either kind of homozygote. Since we do not have the *th/th* homozygote, this plan would not prove effective. However, by selecting for the heterozygotes and for any *th/th* homozygotes produced, we should expect to get our necessary homozygotes. If we let only these two genotypes breed, we are creating a Case I or II environment, and eventually all the offspring will be homozygous *th/th*.
>
> We also know that, through inbreeding, we can increase the frequencies of the homozygotes at the expense of the heterozygotes. So by dividing our breeding chambers in such a way that we assure sibling mating, some of these chambers will eventually have the homozygous *th/th* individuals.

After much trial and error, the proper mix of genotypes is obtained and I set out my samples. I find that the thin-plated homozygotes suffer a high degree of mortality due to crushing by waves. The heterozygotes and the thick-plated homozygotes have essentially the same mortality, but quite a bit less than the *th/th* genotypes. The results are not significantly different at the two sites. I calculate that the selection coefficient against the *th/th* genotype is 0.35. In addition, I find no recordable differences in fecundity from the laboratory experiments. Our quick-and-dirty analysis thus gives fitness values of $W_{Th/Th} = 1.0$, $W_{Th/th} = 1.0$, and $W_{th-th} = 0.65$. Can we begin to eliminate some of our hypotheses for the mechanism of maintaining the polymorphism?

If we assume that our rough analysis of fitness is correct, we can immediately reject some of our hypotheses. The first hypothesis that we may reject is that of overdominance. The fitness values show clear and strong selection against the thin-plated genotype. If selection were acting alone, we would not expect to find a balanced polymorphism in the population. The Th allele should be the only one present. If our populations are in equilibrium as we assumed, selection must be balanced by another evolutionary force.

Could there be a balance between migration and selection? At face value the answer is yes, but let us investigate more closely. First, we need to set up our balanced equilibrium equation for selection and migration. As in Lecture 7, it is

$$\Delta q = 0 = -sq^{*2}p^* + m(q_m - q_h^*)$$

If we assume that the migration is coming from a source population where $q_m = 1$ (that is, the source population is fixed for the th allele), then we can solve for m in site A as follows:

$$0 = -(0.35)(0.001)^2(0.999) + m(1.0 - 0.001)$$

$$= -3.5 \times 10^{-7} + m(0.999)$$

$$m = \frac{3.5 \times 10^{-7}}{0.999}$$

$$m = 3.5 \times 10^{-7}$$

Thus to balance selection, we must expect a 0.00004% migration rate per generation. Considering the pelagic dispersal of the larval stage, this is an absurdly low rate. Of course, if we were to assume that the source population did not have a gene frequency of $q_m = 1.0$, we could come up with more reasonable migration rates, but then we must explain what is maintaining the polymorphism in the source population.

Will a mutation-selection balance provide more reasonable values? Let us try by setting up the equilibrium value of q, assuming selection against the recessive. From Lecture 8, we know this equation to be

$$q^* = \sqrt{\mu/s}$$

For site A, we can solve for μ as follows:

$$\mu_A = q^{*2}s$$

$$= (0.001)^2(0.35)$$

$$= 3.5 \times 10^{-7}$$

This is a reasonable mutation rate. What about at site B?

$$\mu_B = (0.0075)^2(0.35)$$
$$= 1.97 \times 10^{-5}$$

This is a bit larger than the mutation rate at site A. It seems that something at site B is increasing the mutation rate and that the gene locus under investigation is a sensitive measure of that mutagenic disturbance. Perhaps it is the power plant.

The possibility that the *th* locus can be used as an environmental monitor is intriguing. Further sampling in the intake and output channels of the power plant is undertaken. I name these sites C and D (see Figure II.1). This time I collect 4000 individuals from each site. The number of individuals of each genotype and the gene frequencies are given in Table II.1.

TABLE II.1 Census by Genotype and Gene Frequencies at Sites C and D

	Th/Th	Th/th	th/th	f_{Th}	f_{th}
Site C	3940	59	1	0.9924	0.0076
Site D	3903	96	1	0.9877	0.0123

The difference between *th* frequencies at sites C and D is quite large, whereas that between sites B and C is negligible. If all our assumptions are correct, then the mutation rate is the same at sites B and C. What is happening at site D? First, the sites are right next to each other and so we would not expect to see a distance effect as between sites A and B. Second, the barnacles in site D come from the same water as in site C, since the water is flowing through the entire system. The water system is closed so that there is no difference in the chemical makeup of the water. The only difference is in the water temperature. What could be causing the differences in frequency?

Assuming that all our previous conclusions are correct, we still should be dealing with a mutation-selection balance. There is no apparent difference between sites B and C, and barnacles from site D come by site C, so it does not look as if migration plays a part in the problem. Going back to our mutation-selection equilibrium, q can be increased either by increasing μ, the mutation rate (as we suggested occurred between sites A and B), or by decreasing the selection coefficient.

Taking these as my working hypotheses, I set up the following experiments. I establish separate tanks with water taken from sites C and D. Half the tanks are maintained at the temperature found at site C and half are maintained at the temperature found at site D. Barnacles are introduced and maintained for many generations.

The results indicate no effect of the source of water on gene frequencies. However, increased temperature causes a higher mortality in Th/Th homozygotes than in either Th/th heterozygotes or th/th homozygotes. I conclude that the difference in gene frequencies could be explained best by an increase in the mortality of the Th/Th homozygotes, thus reducing the relative selection coefficient against the th/th and Th/th genotypes at site D. Indeed, further physiological investigation indicates that the intracellular pH differences found in the Th/th and th/th genotypes assist in the efficiency of the oxygen metabolism at higher temperatures, the th/th metabolism being slightly more efficient. Given that all these laboratory results reflect what is happening at the field sites, what is the selection coefficient against the th/th homozygote at site D?

> Let us solve this problem assuming for simplicity complete dominance. We will also assume that the mutation rates are the same at sites D and B, that is,
>
> $$\mu_D = 1.98 \times 10^{-5}$$
>
> Using this value and the q^* value, we can solve for the selection coefficient s:
>
> $$s \approx \frac{\mu}{q^{*2}}$$
>
> $$s_D = \frac{1.97 \times 10^{-5}}{(0.0123)^2}$$
>
> $$s_D = 0.13$$
>
> This value is quite a bit less than the suggested selection coefficient value of 0.35 at sites A, B, and C.

From the mapping of the water currents in the area, it is clear that the water masses mix off the jut of land marked site E. If the larvae are taken from sites A and D, mixed in the water column, and settle at site E, we should expect to see a depression in the observed number of heterozygotes due to the Wahlund effect. What is the expected depression in heterozygotes, F_{ST}, and do you really expect to see such a depression if you sample at this site?

> From Lecture 8, the frequency of heterozygotes observed is less than the expected frequency by a factor of two variances, or
>
> $$H_e = H_o + 2\sigma^2$$
>
> $$2\bar{p}\bar{q} = \frac{(2p_1q_1 + 2p_2q_2)}{2} + 2\sigma^2$$

where

$$\sigma^2 = \frac{p_1^2 + p_2^2}{2} - \bar{p}^2$$

The inbreeding coefficient is then

$$F_{ST} = \frac{\sigma^2}{\bar{p}\bar{q}}$$

By using the frequencies of the *th* allele at sites A and D, we can compute F_{ST} as follows. First, we compute the average gene frequencies.

$$\bar{p} = \frac{p_A + p_D}{2}$$

$$\bar{p} = \frac{0.999 + 0.9877}{2} = 0.9933$$

$$\bar{q} = 1 - \bar{p}$$

$$\bar{q} = 0.0066$$

We next determine the variance, using the preceding equation:

$$\sigma^2 = \frac{1}{2}[(0.9990)^2 + (0.9877)^2] - (0.9933)^2$$

$$\sigma^2 = 0.00013$$

The expected depression in the frequency of heterozygotes is two times this number, or 0.00026. The inbreeding coefficient is

$$F = \frac{0.00013}{(0.9933)(0.0066)}$$

$$F = 0.0200$$

Lastly, will we really notice this depression? Here again we will not bother with a formal statistical analysis. Our intuition will serve us perfectly well in this case. Stated another way, how many barnacles must you sample until the expected depression corresponds to one barnacle? To solve this, we divide one by the expected frequency deficit.

$$N = \frac{1}{0.00026}$$

$$N = 3846.2$$

Of this number, we would expect to see

$$2\bar{p}\bar{q}N = 50.42$$

heterozygotes. What if we find only 49? I would not be suspicious of a real deficiency if I saw 49 instead of 50 heterozygotes in 3846 barnacles. It would take a very large sample, larger than you or I would care to take, to become suspicious. Therefore, in the case of our barnacles, where we are dealing with such low gene frequencies and small differences between populations, the Wahlund effect is not noticeable.

Homework exercises

The following should be understood to solve the problems.
For mutation-selection balance, the frequency of the recessive allele q, which is driven out by selection but is constantly produced by mutation from the dominant form, is

$$\boxed{q = \sqrt{\mu/s}}$$

where μ is the mutation rate and s is the fitness deficiency of the recessive homozygote in the fitness scheme $(1, 1, 1 - s)$. A typical error is to confuse s with $(1 - s)$. If, for instance, we have a recessive homozygote fitness of 0.8 as compared to 1.0 for a dominant homozygote and heterozygote, $s = 1 - 0.8 = 0.2$. Do not put 0.8 in the formula instead of 0.2. The other error is more fundamental. Remember q is the allelic frequency and q^2 the genotypic frequency. The question may be asked about the occurrence of the disease, that is, the percentage of sick individuals: They are genotypes, not alleles. The frequency is, therefore, $q^2 = \mu/s$.

In the migration-selection problem remember the balance equation

$$\boxed{-sp_h q_h^2 + m(q_m - q_h) = 0}$$

The problem usually gives information about frequencies and asks for either m (migration rate) or s (selection intensity). The equation is linear in s and in m so that you should have no difficulty finding the necessary value.

For problems on inbreeding remember that the inbreeding coefficient F_{IS} (due to nonrandom mating) is defined as

$$F_{IS} = \frac{H_e - H_o}{H_e}$$

where H_o is the observed proportion of heterozygotes and H_e is the expected proportion of heterozygotes (under the Hardy-Weinberg law). The inbreeding coefficient due to the population subdivision into two separate populations is

$$F_{ST} = \frac{\sigma^2}{\bar{p}\bar{q}}$$

where

$$\bar{p} = \frac{1}{2}(p_1 + p_2)$$

$$\bar{q} = \frac{1}{2}(q_1 + q_2)$$

$$\sigma^2 = \frac{1}{2}[(p_1 - \bar{p})^2 + (p_2 - \bar{p})^2]$$

when the two population sizes are equal, $N_1 = N_2$.

1. The selection coefficient against a deleterious mutation in mice is known to be $s = 0.9$ for the recessive homozygote. If we observe a frequency of 0.9996 of *normal* individuals in a given population, what is the mututation rate μ? (Assume Hardy-Weinberg equilibrium.)

2. A genetic mutation that affects the phenylalanine-tyrosine metabolic pathway causes a disease called phenylketonuria. This disease can cause extreme mental defects. When diagnosed at an early stage, however, dietary precautions can be taken which nullify the harmful effects of the disease. Human populations were sampled from an isolated region of Appalachia and from New York, New York. The gene frequencies of the mutant allele were 1×10^{-5} and 5×10^{-5}, respectively. Assuming no migration, large panmictic populations, and constant mutation rates, what are the relative selection coefficients of the homozygous condition in Appalachia and New York, that is, how much stronger is selection in Appalachia than in New York?

3. Recent evidence indicates that persons living near toxic chemical waste dumps may have a higher incidence of deleterious mutations, as evidenced by an apparent increase in the proportion of children born with birth defects. Suppose we obtain data on birth defects for two groups of people: One lives in a relatively pristine environment in the country, and the other has recently been evacuated from a suburban housing development inadvertently constructed on a landfill polluted with mutagenic chemical wastes. If the frequency of the recessive lethal allele q is 0.001 in the normal population, what is the mutation rate under unpolluted conditions? If we observe a twofold increase in the frequency of affected children born to parents from the polluted landfill, what is the increase in the mutation rate for the exposed population?

4. Children born with another birth defect have a 40% chance of surviving to live and reproduce normally. If the mutation rate for the recessive allele which is responsible for the disease is increased from 1×10^{-5} to 4×10^{-5} among people exposed to the harmful chemical pollutants, what is the expected difference between populations in the frequency of heterozygote carriers of the disease?

5. A field mouse has a color polymorphism that is controlled by two alleles at one locus. In a certain field it was found that $P_{tan} = 0.20$ and $P_{brown} = 0.80$. These frequencies were stable over several generations. After a careful natural history study, it was found that the tan mice were twice as likely as the brown mice to be killed by predatory hawks. There was no difference in fecundity or in any other source of mortality. Careful mark-recapture studies indicated that an average of 4% of the mice were immigrants per generation. Assuming that the polymorphism is maintained by migration, what is the frequency of the tan allele in the source population?

6. The uniformly grayish, unbanded color pattern of water snakes is determined by an allele (B) which is dominant to the recessive banded allele (b). A large mainland population of water snakes is made up almost entirely of banded individuals (bb). On a nearby island, however, the uniform unbanded color pattern is more common. A careful natural history study reveals that there is strong selection on the island against banded snakes. If the frequency of the banded allele (b) on the mainland is $q_m = 0.9$, the frequency of unbanded alleles (B) on the island is $p_h = 0.6$, and the fitness of recessive banded individuals on the island is $W_{bb} = 0.2$, what is the equilibrium migration rate m between the mainland and the island?

7. A population of moths consists of two color morphs, brown (BB and Bb) and white (bb). In a population of 100 moths we observe 91 brown and 9 white individuals.
   (a) If the population is in Hardy-Weinberg equilibrium, what is $2pq$, the frequency of heterozygotes?
   (b) Suppose that we determine that only 28 of the moths are heterozygous. What is F_{IS}, the inbreeding coefficient?

8. The edible Atlantic mussel is known to have two color morphs: brown and blue-black. There is some controversy among taxonomists over the classification of these mollusks, some believing that the two color morphs are a polymorphism within one species (*Mytilus edulis*) and others that the two morphs are actually different species (*M. edulis* and *M. galloprovincialus*). Animals of the two morphs were collected from one site. The gene frequencies of a marker allele determined for each morph are given.

Morph	Gene Frequency
Blue-black	0.7
Brown	0.2

If the two morphs are indeed separate species, we should expect a decrease in heterozygotes when viewed over the pooled collection. What is the expected F_{ST} if this is the case? What is the expected heterozygosity of each population individually?

chapter III

Genetic Drift and Neutral Evolution

lecture 9
Sampling errors

The theory of neutral evolution is an area of population genetics that has advanced with great strides during the past 10 to 15 years. It has reorganized our thinking about the causes of evolution by building a non-Darwinian model for genetic change in populations. Controversy still rages over the extent of neutral evolution in natural populations, but as an important force to be considered it is here to stay.

The material in this lecture will exercise your skill at making connections between one area of activity and another. Specifically, I will start by talking about gambling, or to be more exact, winning when gambling.

The game we will play is "heads or tails." We will use a completely fair coin in which the chance of getting either a heads or tails is exactly one-half. The rules of the game are as follows. Each of us puts in $1 for each throw of the coin. If the tossed coin comes up heads, I win and take the money; if the tossed coin comes up tails, you win and take the money. The game continues until one of us has no more money to play.

Who has a greater probability of winning? Your intuition may tell you that we both have equal chances. The coin is fair, after all, and the return on each toss is the same. You either lose a dollar or win a dollar each time. If this is your conclusion, then your thinking is right but your answer is wrong.

If I were to ask you before we started the game, how much money do you *expect* to have after 20, 30, or 100 tosses, how would you answer? You may say that you cannot even venture a guess; it all depends on how many times you win or lose. To be sure, but this prudent answer is to another question: How much money will you actually have? We can say how much you expect to have by calculating how many tosses you expect to win. If the coin is really fair, you should expect to win 10 in 20, 15 in 30, and 50 in 100. In 20 tosses you will win $10 and lose $10. In 30 tosses you will win $15 and lose $15, and so on for an infinite number of tosses. Therefore, you would expect to have the same amount of money as when you began for as long as the game progresses.

Will the game go on forever with neither of us winning? Just because the coin is fair does not mean that for every heads we will get a tails. It only means that we have the same probability of getting a heads or tails on every toss of the coin. It is not at all surprising to get two heads in a row. This is because each toss is independent of all others, that is, the coin does not remember what happened in the past. The probability of getting two heads and that of getting a heads and then a tails are the same. When we say that we expect to win as many tosses as we lose, this does not mean that we will actually do so. There is only a certain probability that this will occur. This probability will be greater than the probability of winning exactly x dollars or the probability of losing exactly x dollars, but, depending on how long you have been playing, it may be small in itself. Do not make a side bet that you will be exactly even after 100 tosses!

You may get a better picture of what is happening during our game if I plot out a probability diagram of the game results. These would be the probabilities of winning exactly $3, losing $5, and so on. On reflection you should immediately realize that these probabilities will depend on how long we play the game. For example, there is a zero probability of winning $20 if we have only played 10 tosses. Similarly, the probability of winning $10 with 10 tosses is much smaller than the probability of winning $10 with 20 tosses. We can write the exact probability of any particular outcome for any particular number of tosses:

$$p = \frac{n!}{a!(n-a)!}\left(\frac{1}{2}\right)^a\left(\frac{1}{2}\right)^{n-a}$$

where n is the number of tosses, a is the number of occurrences we want, such as the number of wins or the number of losses, and $1/2$ is both the probability of getting a heads or tails. Clearly, $(1/2)^a(1/2)^{n-a}$ is the probability of exactly (a) heads and $(n-a)$ tails in any specific order. However there are $n!/a!(n-a)!$ ways in which (a) tails can occur. This is the particular formula when we have exactly one-half probability of each outcome. If the two outcomes are not equally probable, that is, one has a probability of p and the other q ($p + q = 1$), the formula can be generalized to

$$p = \frac{n!}{a!(n-a)!} p^a q^{n-a}$$

For example, the exact probability of winning $10 in 20 tosses is the probability of winning 15 tosses and losing only five, which is

$$p = \frac{20!}{15!5!}\left(\frac{1}{2}\right)^{15}\left(\frac{1}{2}\right)^{5}$$
$$= 0.015$$

Figure 9.1 shows the distribution of probabilities for all possible wins and losses in 20 tosses of the coin.

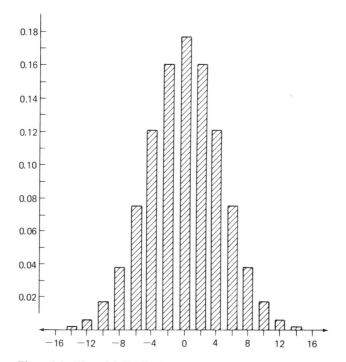

Figure 9.1 Binomial distribution based on 20 tosses of a fair coin.

This distribution of probabilities is called a *binomial distribution*. If you are unfamiliar with this, do not worry. There are only a few simple points that we must understand from this formula. First, the probability of losing a certain amount is exactly equal to winning the same amount. You can see this by noticing that both win and loss terms are included in the numerator, so that the probabilities are symmetric. To put this even more simply, both you and I have the exact same chance to win $50. This is another way of saying that the game is fair.

The second point is more involved and is counterintuitive; it is also essential to our further understanding. It seems logical that, while we may deviate from the balance of equal wins and losses, the longer we play, the more these deviations will even out and the closer we will be to our starting point. Unfortunately, this is not true. The longer we play, the smaller the chance that we will be at the center of distribution and the greater the chance that we will be at one or the other extreme. We will either win a lot or lose a lot, the more we play. In Table 9.1 I have listed the probabilities of winning or losing $10 or less and of winning or losing more than $10 for different numbers of coin tosses. The general trend is clear.

TABLE 9.1 Probabilities of Results

Tosses	Winning or Losing $10 or Less	Winning or Losing More Than $10
50	0.88	0.12
100	0.73	0.27
200	0.48	0.52
500	0.34	0.66
1000	0.25	0.75

There are two ways to understand this result. Consider first that, as the number of tosses increases, the probabilities of greater individual gains or losses increase, as in our previous example where the probability of winning $10 in 10 tosses is less than the probability of winning $10 in 20 tosses. The area under the tails of our probability distribution thus increases with the number of games. Second, remember that each toss effectively starts a new game. There will be no more tendency to get a heads after getting 10 tails in a row than there was before the 10 tails. If you have lost $10, you have a greater chance of losing $5 more than of winning back your original $10.

What does all this have to do with our game? Let us say that you start out with $5 and I with $20. I will plot our combined gaming resources on one line as shown in Figure 9.2. I will superimpose Figure 9.1 on Figure 9.2 so that the zero point is exactly at our starting positions, that is, at the $5 mark. Now we ask, what is the probability that you will win all my money versus the probability that I will drive your resources to zero? This is the probability of your getting +20 or more versus −5 or more. Obviously, the latter probability is greater. Consider a marathon heads-and-tails session, where we play thousands of games, each one ending when one of us has all the money and each one beginning at the same starting conditions of monetary reserves. You will win some of those games, but I will win the majority. The proportion of games that you can expect to win, or the probability of your winning a particular game, is equal to the proportion of the entire sum of money in that

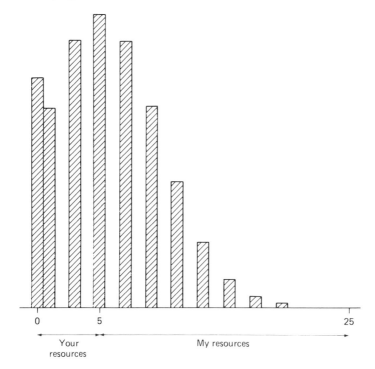

Figure 9.2 Probability distribution of the possible outcomes of coin toss game after 20 tosses.

game that you start with. In the example you had $5 out of the total $25 in the game. Therefore, the probability of your winning a particular game is

$$p(\text{your winning}) = \frac{5}{25} = 0.20$$

This general conclusion is known as *gambler's ruin*, which implies that, even with a totally fair game, the chances of coming away a winner depends on your reserves relative to those of your opponent. Thus professional casinos must maintain large sums of money to help insure their chances of making a profit. Although the usual odds scheme is slightly weighted in their favor in most games, they are not dependent on "fixed" games to make money. Having a large reserve, they will win even if all the games are fair. The moral of this lecture is self-evident, but like most such things, different morals are self-evident to different people. You must decide whether the moral is not to gamble or not to gamble with people richer than yourself.

Enlightening as this may be, what does it have to do with genetics? If you will forgive me for anthropomorphizing again, every allele is "gambling" each time a population reproduces. The tosses of the coin are the gametes that fuse to form fertile zygotes. They either carry the gene or they do not. The number of tosses, then, is the number of gametes that are included in syngamy. In a diploid population with N individuals this number is $2N$. The only difference between this game and our own is that this is the ultimate existential gamble: If you lose, you cease to exist.

We will start again with the equilibrium Hardy-Weinberg population. Let us accept all the assumptions that we made except that we will now consider populations that are not necessarily large. The description "not necessarily large" is a general statement, but I do not want to get more specific at present.

Assume that each individual releases the same number of gametes so that the allelic frequencies in the population of gametes is equal to the allelic frequencies in the parental population. From this pool of gametes, $2N$ are chosen to form the progeny population consisting of N individuals. The allelic frequencies in the progeny population should be equal to those of the gamete population since the gametes were selected randomly. As in our gambling when we spoke of wins or losses, the emphasis must be placed on the word "expect." In exactly the same manner an allele will have an expected loss or gain of zero each generation. The expected value of p' is p each generation. In all actuality, however, p' will probably not be equal to p in each subsequent generation. It will either gain or lose randomly as in our coin toss game.

This random increase or decrease in frequency of an allele is due to *sampling errors*. As one particular toss of the coin cannot be half heads and half tails, a gamete must carry either a or b alleles and not a proportion p of a and q of b. Therefore, as long as there is a finite number of sampling events (either coin tosses or sampling of gametes), there will probably be some deviation from the expected frequencies. This sampling deviation in gene frequencies in natural populations is called *genetic drift*.

We return to the general subject of evolutionary change. If there is deviation in gene frequencies due to sampling error, then there is some Δp or evolution occurring in our population. How large is Δp, or alternatively, how strong an evolutionary force is genetic drift? It is impossible to answer this in a totally direct manner as it was impossible to resolve how much money you will win in 30 tosses. As stated, the expected Δp will be zero, but there will be nonzero deviations occurring with certain probabilities.

We must return to the binomial distribution to attack this problem. Since any particular gain or loss of copies of alleles due to genetic drift has only a small probability, it is better to study a range of possible deviations that together will cover a large proportion of the possible deviations. Such a range can be defined by the *standard deviation* of the distribution. A standard deviation of a distribution is the square root of the variance of that distribution and is denoted by σ.

When variance was introduced in Lecture 4, I mentioned that it is a measure of the spread of a distribution around its mean. This should be clearer to you now. I will take the expected value of the frequency of the a allele, which is p, and describe a range of frequencies from $p - \sigma$ to $p + \sigma$. The sum of the probabilities within our range will be 68.26%.* Furthermore, a range of two units of standard deviation on either side of the mean will account for 95.46% of all probable deviations. Using this type of argument, we can decide what is a likely or, more importantly, what is an unlikely deviation, Δp, due to genetic drift.

To make such decisions, we must decide what we mean by likely and unlikely and what we can determine to be our standard deviation. It is best to employ the standard scientific limits to likely and unlikely events. The basic argument is as follows. I take a measurement or a sample of something (allelic frequencies in our case) and get a deviation from the expected. For all the reasons that we discussed, a small deviation is not surprising. However, a very large deviation due to sampling error alone *is* surprising. The general convention used is that, if the expected probability of obtaining a deviation as large or larger than that is 5% or less, we assume that it is unlikely that the deviation is due to the supposed mechanism. Some other processes may be involved; in our case, we would infer that other forces besides drift have acted.

Note two important concepts:

1. Even though there is a less than 5% chance of deviation this large or larger occurring due to sampling error alone, and we therefore infer that other forces are involved, we are not positive that this deviation is *not due only to drift*. It is just determined to be unlikely.
2. If there is a greater than 5% chance of deviations this large or larger being due to drift, we are not positive that this deviation *is due to drift*. It could be, or it could be due to other forces. We cannot conclude anything definite in such a case.

We must now determine the standard deviation of our distribution. The binomial distribution has a simple expression for the standard deviation,

$$\sigma = \sqrt{pq/2N}$$

where p and q are the allelic frequencies and $2N$ is the number of gametes sampled. For our decision-making process, we wish to describe the range that covers about 95% of the probabilities. This is equal to the range $p - 2\sigma$ to $p + 2\sigma$. If our deviation is outside this range, we assume that our Δp is not due to genetic drift.

We should be aware of two important characteristics of genetic drift. The standard deviation that describes the expected distribution of deviations or Δp's due to sampling error is

$$\sigma = \sqrt{pq/2N}$$

*The probabilities listed here are derived from the normal distribution and are not precisely correct for the binomial distribution. However, it can be demonstrated that, as N becomes large, the binomial distribution quickly approaches normal. As a result, I have been using the common decision-making rule of thumb based on the normal distribution.

Notice that as N increases, σ will decrease. In the original assumptions of a Hardy-Weinberg population the population was large. It should be apparent to you that drift will be an insignificant force if N is large. For example, if $p = 0.20$, $q = 0.80$, and $N = 80$, $\sigma \approx 0.03$. Hence, 95% of all deviations will be in the range $p' = 0.14$ and $p' = 0.26$. On the other hand, if $N = 10,000$ and $\sigma = 0.003$, 95% of the deviations will be in the range between $p' = 0.194$ and $p' = 0.206$. For a population size of 1 million, we would never notice the change.

Consider how drift is influenced by frequencies. By keeping N constant, we can maximize σ by maximizing the numerator pq. You should be able to convince yourself that pq is maximized when p is 0.5. Similarly, σ decreases as p or q goes to one (Figure 9.3). This means that drift will have its greatest effect on populations with intermediate gene frequencies. For instance, if $N = 50$ and $p = q = 0.5$, the frequency range of two standard deviations will be $p = 0.5 \pm 0.1$. If, however, $p = 0.95$ and $q = 0.05$, the range will be $p = 0.95 \pm 0.04$. It is harder to drift away from the ends of the frequency spectrum than from the middle.

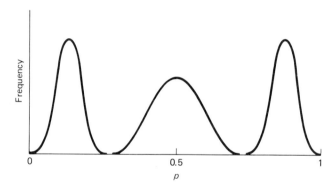

Figure 9.3 Binomial distributions, indicating the effect of initial frequency on the standard deviation.

Let us quickly review this lecture. Random sampling will result in deviations from the expected value. Therefore, even if there are no deterministic evolutionary forces acting on a population (such as natural selection), sampling error during the production of zygotes may cause p' to be unequal to p. This deviation in gene frequencies from one generation to the next is called genetic drift. We can set limits to the extent that drift is likely to cause changes by determining if p' is within the range of $p - 2\sigma$ and $p + 2\sigma$, where the standard deviation is $\sqrt{pq/2N}$. If p' is outside this range, it is unlikely that drift is the cause. If p' is within this range, Δp may be due to drift. In general genetic drift will be strongest when population size is small and gene frequencies are intermediate. Finally, if it is ever prudent to gamble, it is only when you are rich.

Lecture 9 Sampling Errors

In the models concerning selection we determined p for the first generation and then derived the ultimate equilibrium gene frequencies, given the initial conditions. We will continue to focus our interest on the ultimate equilibrium conditions. What are the equilibrium conditions under genetic drift? How strong is genetic drift compared to other evolutionary forces? Lecture 10 will address these problems.

lecture 10
Drift
in small populations

In Lecture 9 we discussed the fate of an allele in a small population. Because of sampling error, the frequency of an allele will change from one generation to the next. We cannot predict whether the allele is likely to increase or decrease at any given time, nor can we say by how much. We can determine, however, the probable range of changes in frequency. Because we are dealing with a random process, we use a 95% probable range. If the absolute value of Δp is greater than $2\sqrt{pq/2N}$, where $\sqrt{pq/2N}$ is the standard deviation of random fluctuations in gene frequencies, then there is only less than a 5% chance that this change is due to drift alone. In such a case we would suspect that other evolutionary forces are acting. If $|\Delta p|$ is within the $2\sqrt{pq/2N}$ limit, we will accept drift as a possible cause of the change.

I introduced the concept of gambler's ruin in the previous lecture. We did not get a chance to develop fully its genetic implications. This lecture will be based on the development of gambler's ruin in the context of many small, isolated subpopulations. Each individual subpopulation is in effect one independent game, and the collection of all subpopulations is the genetic analogue to our coin-tossing marathon.

We will begin with one large population in which the frequencies of the a and b alleles are p and q, respectively, as usual. The alleles are distributed evenly throughout the entire population so that, when we break up this large population into isolated groups, each group has an identical gene frequency, p and q. The distribution of gene frequencies is a spike, showing that 100% of the subpopulations have a frequency of a at p (Figure 10.1). Each subpopulation then undergoes one generation of random mating. Although on the

Lecture 10   Drift in Small Populations   91

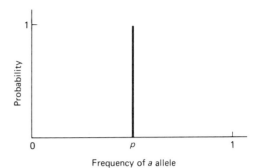

Figure 10.1 Distribution of gene frequencies of subpopulations at the beginning of the experiment.

average each subpopulation should retain the same gene frequency p, in reality each of them deviates to some extent because of genetic drift. If I plot the distribution of gene frequencies after one generation, I would no longer have a spike, but rather a bell-shaped curve (Figure 10.2). Most populations will deviate only slightly from the original gene frequency, but a few will have large deviations. If I let the process continue for several generations, the gene frequencies will deviate further until the frequency distribution is essentially flat (Figure 10.3). At this point, any nonzero or nonone frequency will be equally likely to occur.

This is an odd and nonintuitive conclusion. To understand fully this distribution pattern, let us again consider just one population. Figure 10.4 shows the distribution of gene frequencies in one population as the frequency

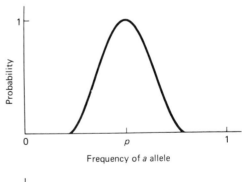

Figure 10.2 Distribution of gene frequencies of subpopulations after one generation.

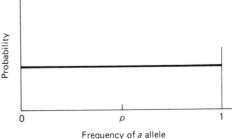

Figure 10.3 An even or rectangular distribution of gene frequencies.

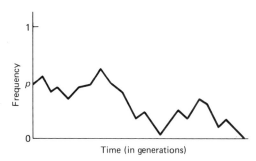

Figure 10.4 Trajectory of a gene frequency as it drifts over time.

drifts over time. The trajectory that the frequencies follow is not a simple straight path, but is rather like a drunkard's meanderings. Because the direction of frequency change can change each generation, a population can potentially reach any gene frequency until the frequencies reach either zero or one. Therefore, if we have many subpopulations, each equally inebriated, we will expect to see an even distribution of gene frequencies.

Why will this wandering continue until the frequencies reach either zero or one? The answer should be obvious. When the frequencies reach zero or one, there is no longer any possibility of increasing the frequency of an allele that has disappeared from the population. Think of the boundaries zero and one as sponges sucking up populations. As time continues, more and more populations get absorbed into the boundaries, but the distribution of the intermediate frequencies remains flat. The "depth" of the intermediate frequencies simply gets shallower and shallower until all the populations are absorbed at the boundaries. The frequency distribution of p and q is then spiked at zero and one. Of course, the rate at which the surfaces sink is dependent on the number of individuals per subpopulation. The lower the population size, the more the allelic frequencies will drift in a generation and the faster they will be absorbed at the boundaries. We encountered this in the last lecture, where the standard deviation of drift increased with decreasing N. The actual rate of sinking is $-1/2N$ (Figure 10.5).

The relevance of this discussion to the general subject of evolutionary forces, developed in the previous two chapters, should be seen through the following example. Perhaps, by so doing, you will get a clearer notion of the types of problems that field population geneticists have to confront.

In Lecture 1 we introduced wild fruit flies into two identical cages. After many generations, one cage had only red-eyed and the other had only white-eyed individuals. Why did the gene frequencies of the two cages diverge so radically? In Lecture 5 we advanced one possible explanation: The system was underdominant and the original gene frequencies were on either side of the p^* value. Now we can pose another possible explanation. The flies did not differ at all in their fitnesses. Because the populations were small, they drifted in opposite directions by pure chance until they became fixed for alternate red-eyed and white-eyed alleles.

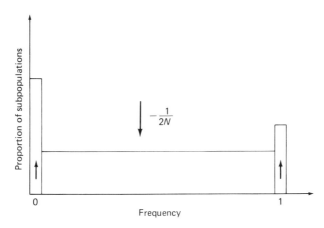

Figure 10.5 Distribution of gene frequencies, showing the piling up of populations with frequencies of zero and one as they are absorbed by the boundaries.

Having two answers can be as bad as having no answers. How do you distinguish between the two? This is not at all trivial. For example, assume that the system is underdominant and that the fitnesses of the genotypes are $W_{rr} = 1.00$, $W_{Rr} = 0.95$, and $W_{RR} = 1.00$. If you investigated just the homozygous individuals, you would find that they have equal fitnesses, and you might conclude that neutral genetic drift occurred. On the other hand, if you tried to investigate all the genotypic fitnesses, you might not be able to determine the real genetic component of the differences because of random environmental effects that could camouflage the difference. Indeed, you would not expect to see large differences in fitnesses simply because large fitness differences would cause rapid changes in gene frequencies. Therefore, in cases of potentially small selection coefficients, you would want a more controlled environment to test for fitness differences. Yet if you worked in a controlled environment, you could not say for sure that the fitness coefficients that you determined are the same as in the original populations. Genotypes react differently to different environments, and this can cause differences in fitness components.

Ascribing population differentiation to drift alone is also a tricky business. Essentially the problem here lies in the parameter N for population size. We have discussed that the amount of drift per generation, and the rate of fixation of alleles, is dependent on N. The larger the population, the slower the drift. Of course, given any population of any size, you can postulate that drift has caused fixation. There is obviously some probability that a large population will drift quickly, but that probability is small. We are left supposing that the populations drifted slowly in opposite directions for many generations. Yet now we have another problem: These populations must have been in complete isolation for all that period of time. It takes little migration to counteract drift. Effectively, migration increases the size of the population and drastically slows down the process.

In addition there is the problem of changes in population size over time. For example, let us observe a population of 10,000 individuals. We would expect a relatively slow rate of drift in a population of this size. Suppose that this population actually started out three generations ago from only 50 individuals. Obviously, we are not accounting for the true potential of drift that our population has had over time if we look only at present population size.

This will be clearer if we go back to the concept of inbreeding coefficients, introduced in Lecture 8. First, remember that the inbreeding coefficient F measures the deficiency of heterozygotes in a population. If all that we have been saying about populations drifting to fixation for either allele is true, then we should expect to see F increase over time due to drift. This can happen in the following way. Suppose we have an effectively infinite pool of gametes and randomly choose pairs to form our zygotes. If there are $2N$ different kinds of gametes, then the probability of getting a pair that is exactly the same, that is, an inbred pair, is $1/2N$. Of those that are not exactly the same $(2N - 1/2N)$, the proportion of inbred pairs will be determined by the inbreeding coefficient of the population before mating, or F_{t-1}. We may write our F coefficient due to sampling error, or drift, as

$$F_t = \frac{1}{2N} + \left(\frac{2N-1}{2N}\right) F_{t-1}$$

If any previous generation had few individuals, the contribution of that generation to future inbreeding coefficients is large. You can see this in the example given in Table 10.1. The population size was stable for six generations except for a drastic drop which occurred in generation 4. After six generations, the average increment of F is 0.0021 (0.0125/6).

TABLE 10.1 Generation of Inbreeding by Drift over Time

t	0	1	2	3	4	5	6
N		1000	1000	1000	50	1000	1000
F	0	0.0005	0.0010	0.0015	0.0115	0.0120	0.0125

This increment should correspond approximately to $1/2N_{ave}$, where N_{ave} is an average population size. This makes good intuitive sense. The arithmetic average of the population size is 842 and $1/2(842) = 0.0006$. This is much too low. Working backward, we get

$$\frac{1}{2N_{ave}} = 0.0021$$

$$N_{ave} = \frac{1}{2(0.0021)} = 240$$

Lecture 10   Drift in Small Populations

This is quite a bit smaller than our previous average. The answer to this problem is that 240 is the harmonic mean of the population size. The harmonic mean of a group of n numbers $x_1, \ldots, x_n$ is defined as

$$\bar{X}_{\text{harm}} = \frac{1}{\frac{1}{n}\sum_{i=1}^{n}\frac{1}{X_i}}$$

Indeed, when we calculate the average

$$N_{\text{ave}} = \left[\frac{1}{6}\left(\frac{1}{1000} + \frac{1}{1000} + \frac{1}{1000} + \frac{1}{50} + \frac{1}{1000} + \frac{1}{1000}\right)\right]^{-1}$$

$$= \left[\frac{1}{6}(0.025)\right]^{-1}$$

$$= [0.0042]^{-1}$$

$$= 240$$

we see that this works out. A harmonic mean differs from an arithmetic mean in that low values reduce the mean more than high values increase it. Therefore, one small population size in the past can have a large effect on present gene frequencies.

All that has been said so far is that drift drives populations toward fixation. When there is genetic variability present, you may say that other forces are responsible for its maintenance. After all, variability disappears under drift. Again, this is astute reasoning, but it is unfortunately wrong. The flaw is a simple one. Variability can be maintained in an equilibrium population by overdominance, but who said that our populations are always in equilibrium? If semiisolated populations are drifting to fixation, but a small amount of migration and mutation is occurring, these subpopulations can maintain polymorphisms.

I have brought the water to you to drink only to have you find, like Tantalus, that it recedes from your lips when you try to drink. There are, however, many predictions that can be made and tested. It is hoped that previous problem-solving sessions have given you a taste of such procedures for the deterministic forces. The theory of the stochastic force, drift, also generates predictions that can be tested. Indeed, a school of thought led primarily by the eminent Japanese geneticist, Motoo Kimura, has developed a whole body of theory since the 1960s dealing with such predictions. This theory, generally discussed under the name *neutral theory*, is the subject of Lecture 11.

lecture 11

Neutral evolution

The neutral theory of evolution, as a counter to the selectionist view, has created much furor and controversy over the past two decades. Population geneticists have often cast caution to the wind and grabbed the standard of one or another school, proclaiming it to be the true model of evolution. As is usually the case, the truth is probably somewhere in the middle. Before discussing the neutralist theory, let us review briefly the history of evolutionary thought.

After the work of the first third of this century, two views of the genome of organisms became prevalent. One saw the genome as a highly integrated and homogeneous mechanism. Most loci are homozygous and the rare heterozygous loci show dominance for the common allele. The rare alleles, generated by mutation, are always deleterious. Natural selection constantly rakes out these harmful alleles and purifies the wild type genome. The other school of thought saw the genome as being highly heterogeneous. In this view overdominant heterozygotes are common and natural selection maintains the variation in the genome.

In the late 1960s, population geneticists began using a technique, electrophoresis, that allowed a relatively quick investigation of genetic variation by detecting the enzyme variants produced by different alleles. Electrophoresis gives a somewhat biased picture because it neither looks at the entire genome, or even a random portion of the genome, nor distinguishes between all the alleles present. Even so, it has allowed a rapid survey of populations of a vast number of organisms. The results have conclusively shown that heterozygous loci are very common.

You would imagine that this finding would end the discussion. Heterozygotes are too common, and the allele frequencies are intermediate and too high to be maintained by a selection-mutation balance. On the other hand, the heterozygotes are almost too common to be accounted for by balancing selection. Of course, directional selection with migration could also maintain variability in populations, as we saw in Lecture 7. Still, about a third of all loci studied are polymorphic. Are all those different alleles really contributing to substantial fitness differences? Do individuals with the different ABO blood groups produce different numbers of viable offspring? Can all the variability be maintained through other mechanisms?

It was in light of such empirical evidence that Motoo Kimura, Masatoshi Nei, James Crow, and others developed the *neutral theory of evolution*. The theory deals specifically with evolution on the molecular and biochemical levels and does not treat morphological evolution directly. Essentially, the theory is based on a few observations that are consistent among many diverse lines of organisms. The first is that point mutations, in which nucleotide bases are substituted in the DNA of a gene, occur randomly throughout the genome. The second is that the rate of change of DNA is high; for example, in mammals, one nucleotide base is substituted in a genome every 2 years. From these observations, it is concluded that the majority of nucleotide substitutions in evolution is the result of random fixation of neutral or almost neutral mutants. Thus the observed polymorphisms are neutral and maintained by a balance of mutation and extinction. Figure 11.1 illustrates the fate of such new alleles that enter a population by mutation and increase or decrease in frequency by random drift. By explaining genetic observations in light of only two parameters, mutation rate and population size, neutral theory provides a null hypothesis. The burden of proof then is moved to those who wish to explain these observations by other evolutionary forces.

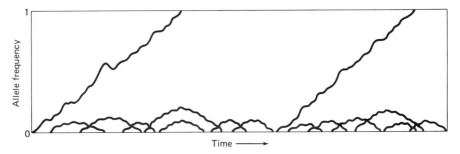

Figure 11.1 The fate of new mutant alleles that are maintained or lost in a population by drift.

You should not get the wrong impression of what the theory states or does not state. First, while the theory deals with neutral nucleotide substitutions, it does not propose that all mutations are therefore neutral. The vast majority of mutations are considered to be highly deleterious or lethal; they never become established in the population. In other words, the neutral theory accepts selection against alleles as does a pure selectionist model. A substitution is simply an eventual fixing of one allele arising from a random mutation in a population.

The second common misconception of the neutral theory is that two gene products, for example, two forms of an enzyme produced by different alleles, have to be identical biochemically or physiologically for the genes to be neutral. Neutrality concerns fitness, that is, the number of viable offspring a genotype will produce. It does not concern enzyme function. Therefore, demonstrations of different catalytic capabilities of the two gene products do not demonstrate different fitnesses of the genes. Organisms are complex mechanisms in which internal, homeostatic controls can compensate for variations in different biochemical pathways.

This is only the tip of the iceberg of differences between a selectionist view and a neutralist view. The body of differences stems from the large amounts of polymorphisms that are found in natural populations. The neutralists contend that these polymorphisms are found because of a balance of mutation and extinction. The selectionists contend that these polymorphisms are maintained by balancing selection. Thus the essential difference between the two theories does not involve the concept of selection itself, but is instead concerned with whether it is only negative and purifying (neutralist) or whether it can also be positive (selectionist).

Be aware that this difference is even more arcane than it may appear at first glance. What if a new allele arose through random mutation and conferred a definite selective advantage to the organism? This would be Case I or II in our selection models. The new allele would rapidly become fixed in the population. Clearly, this would be a case for positive selection. A dyed-in-the-wool neutralist, however, would assert that this is not positive selection for the new allele, but is merely negative selection against the old allele. So the real difference in outlook can be distinguished only by examining polymorphic loci.

Neutral theory is not just an intellectual gadfly. There are predictions, many of them elegantly simple, that can be made and tested. Let us begin with a population of N diploid individuals. This population has a mutation rate μ of nondeleterious alleles per gamete in a generation. If there are $2N$ gametes in a population and the mutation rate is μ, the number of new mutants in the population per generation is

$$2N\mu$$

I will temporarily denote the probability that an allele will become fixed as ω. So the rate at which a new allele will be substituted in the population, K, will be

$$K = \text{(number of new alleles)} \times \text{(probability that an allele will be fixed)}$$
$$= (2N\mu)(\omega)$$

Lecture 11   Neutral Evolution

We can simplify this by using the conclusion from the gambler's ruin. The probability that an allele will become fixed is equal to its initial frequency p. The frequency of a new mutant appearing in only one gamete out of the total $2N$ gametes in the population is

$$p = \frac{1}{2N}$$

Thus we know that

$$\omega = p = \frac{1}{2N}$$

and

$$K = (2N\mu)\left(\frac{1}{2N}\right)$$
$$= \mu$$

The rate of allele substitution is solely dependent on mutation rate μ.

This is a striking conclusion and the basis of a theory commonly referred to as *the molecular clock*. Simply stated, if mutation is a random and constant process, then the number of nucleotide substitutions between two species should be directly correlated to the length of time since the two species diverged. If selection is driving nucleotide substitution, then the probability of fixation of an allele can be shown to equal $2s$, where s is the selective advantage of that allele. The rate of substitution is then

$$K = 2N\mu \times 2s$$
$$K = 4Ns\mu$$

and is dependent not only on mutation rate, but also on population size and selective advantage. These obviously will not be constant over time. When data are plotted out, such as in the case of seven proteins shown in Figure 11.2, the line describing substitutions as a function of time from divergence is surprisingly straight. Furthermore, when comparing parts of molecules that have little or no function to parts that are highly specific in function, the observed number of mutations is higher in the nonfunctional sections, presumably because purifying selection is less intense. Selectionists, of course, do not find anything incompatible in the latter observation, but argue that the constant rates of substitution of the former observation are the result of averaging the rates over time. They argue that the rates of substitution vary with selection, but are ultimately limited by the rates of mutation. Thus essentially both the selectionist and neutralist models are ultimately limited by the process of mutation.

What about polymorphisms? This, as mentioned before, is the essence of the argument. The level of heterozygosity that is predicted by neutralist theory is

$$H = \frac{4N\mu}{4N\mu + 1}$$

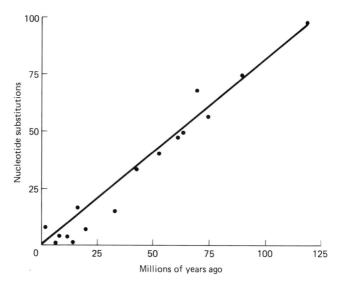

Figure 11.2 Estimated nucleotide substitutions in seven proteins plotted over time in 16 pairs of mammals.

where N is the size of the population of breeding individuals and μ is the mutation rate. (The derivation of this formula is too technical to be included in this lecture.) The level of heterozygosity expected by selectionist theory is, of course, dependent on selective pressures, and about any level of heterozygosity may be explained by adjusting the selection coefficients. We have already stated at the end of Lecture 5 that this robustness of selection theory is its own weakness. (Remember the philosophical problem of the Ptolemaic models discussed in Lecture 1?) On the other hand, the neutralist theory is weakened by its specificity. For example, if the average heterozygosity of *Drosophila* is 0.145 and the heterozygosity for humans is 0.067, we can compare values of $N\mu$. First, we set up the expected heterozygosities and solve for $N\mu$ in each case.

$$H_{Drosophila} = \frac{4N\mu}{4N\mu + 1} = 0.145$$

$$4N\mu = 0.145 + 4N\mu(0.145)$$

$$(4 - 0.145)N\mu = 0.145$$

$$N\mu = \frac{0.145}{3.855} = 0.038$$

$$H_{H.\,sapiens} = \frac{4N\mu}{4N\mu + 1} = 0.067$$

$$4N\mu = 0.067 + 4N\mu(0.067)$$

$$(4 - 0.067)N\mu = 0.067$$

$$N\mu = \frac{0.067}{3.933} = 0.017$$

If the mutation rates are independent of species, we can assume that μ is the same in *Drosophila* and humans. Then the ratio of population sizes of *Drosophila* and humans is

$$\frac{N_{Drosophila}}{N_{H.\,sapiens}} = \frac{0.038}{0.017} = 2.23$$

According to this, *Drosophila* populations are only twice as large as human populations, which seems unreasonable. A neutralist rejoinder would be that N is the breeding size of a population and is greatly affected by small population sizes in the past, as discussed in Lecture 10. The arguments never end.

We are left in a quandary: We have a dichotomy of theories from which to choose. Only difficult field observations and time will sort out the differences. But, as with most dichotomies, the distinction between the two theories may be artificial. Why should populations evolve by only one process or the other? Most likely, some traits, like possibly the precise molecular configuration of cytochrome c, do evolve neutrally and others, like alcohol dehydrogenase or the elephant's trunk, evolve by selection. How can you decide when one or the other is the case? Use your intuition!

Problem solving

In the last problem session it was mentioned that barnacles have a long dispersal stage and therefore can have appreciable gene flow between populations. They can also have large population sizes. We would not expect that genetic drift would be a likely force in barnacle evolution. To look at populations where genetic drift may be occurring, let us go to an alpine environment at the top of a ridge of mountains where we will look at populations of butterflies.

The behavior and habitat requirements of these butterflies make them ideal subjects for the study of isolated populations. The butterflies are extremely heat sensitive. They usually die after only several hours in the lower altitudes found between mountains. As a result, mountain populations of these butterflies are almost completely isolated from each other, except for chance migration events such as when an individual is blown to a new population by a storm.

We will focus on the color pigment found in the wings of the insect. One morph has orange wings and the other white. The color morphs follow simple Mendelian inheritance, where white is dominant over orange. The basis of the color morphs is an enzyme which inactivates the orange pigment and produces a colorless pigment molecule. In the orange morph this enzyme is not active.

The first population investigated is a large, stable population of butterflies. There are well over 5000 individuals in the population by my estimation. I take a random sample of the population and determine that there are 83% white individuals and 17% orange individuals. Assuming that the population is in Hardy-Weinberg equilibrium, these phenotypic frequencies correspond to allele frequencies of $p = 0.59$ for the white-winged or active enzyme allele and $q = 0.41$ for the orange-winged or nonactive enzyme allele. What could be maintaining the color polymorphism?

Without going into in-depth studies on the life history of the butterflies, we can logically eliminate several mechanisms. We will do this by discussing briefly, in checklist fashion, all the potential mechanisms that we have learned to date.

Simple selection with constant fitnesses can maintain polymorphisms if the selection regime is overdominant, that is, if there is heterozygotic superiority. Very high frequencies, around 0.50 for both alleles, can be maintained under either strong or weak selection pressures as long as the homozygotes are roughly equally disadvantaged in relation to the fitness of the heterozygote. My first reaction is to reject this possibility since it appears that the homozygous-white and the heterozygote are phenotypically identical and should have the same fitness. Of course, there could be secondary differences between them that are not immediately apparent which could cause fitness differentials. Thus we are uncertain as to whether or not overdominance is maintaining the polymorphism.

A selection-migration balance could maintain the polymorphism if either the orange or white morphs were selected against and that same morph constantly migrated into the population. Intermediate allele frequencies could be attained with reasonable migration rates, depending on the magnitude of the selection coefficients. Yet if the life history and behavioral information about the butterflies are correct, migration is a rare event. We must therefore reject selection-migration as a reasonable hypothesis.

What about mutation-selection balance? We can almost reject this possibility out of hand. It would take extremely high mutation rates or extremely low selection coefficients to reach such equilibrium frequencies. Such high mutation rates probably do not exist. Such low selection coefficients would be so weak a force that other forces, such as drift, would become more important.

Genetic drift is our last possibility. We assume that the polymorphism is simply a nonequilibrium point in time and that the population is drifting very slowly. If the census data that I estimated are correct and if the population maintained this large size in the past, then this is a reasonable hypothesis. Drift could be acting on this population of butterflies.

Fortuitously, a sudden, freak gale occurs and blows butterflies to two neighboring, previously uninhabited mountain tops. On the first mountain only 12 individuals, six males and six females, become established. The frequency of the white allele in this population is only $p = 0.45$. On the second mountain 294 individuals become established, and the white allelic frequency is $p = 0.62$. Is there any evidence that selection occurred in the establishment of these two new populations?

Let me first rephrase the question. Are the deviations in gene frequencies that we see in the two populations within the range of expected deviation due to sampling error? If they are, then there is no direct evidence from the frequencies alone that the changes in gene frequency are due to selection. In the same breath if the deviations are in the range of sampling error, this does not mean that selection did not cause the changes.

Remember from Lecture 9 that approximately 95% of all sampling deviations will be in the range of $\pm 2\sigma$, where $\sigma = \sqrt{pq/2N}$. The observed deviations Δp from the original population are

$$\Delta p_1 = p_1 - p_{\text{original}}$$
$$= 0.45 - 0.59 = -0.14$$
$$\Delta p_2 = p_2 - p_{\text{original}}$$
$$= 0.62 - 0.59 = 0.03$$

All we have to do is to compare 2σ for each population to the observed deviations.

$$2\sigma_1 = 2\left(\sqrt{\frac{p_o q_o}{2N_1}}\right) \qquad 2\sigma_2 = 2\left(\sqrt{\frac{p_o q_o}{2N_2}}\right)$$
$$= 2\left(\sqrt{\frac{(0.59)(0.41)}{2(12)}}\right) \qquad = 2\left(\sqrt{\frac{(0.59)(0.41)}{2(294)}}\right)$$
$$= 0.20 \qquad = 0.04$$

Notice that we use the gene frequencies of the original population, not of the two new populations. We do this because the sampling event, which in our case was the gale winds, acted on the original population.

Because both deviations are within the range of $\pm 2\sigma$, the deviations may be due to sampling error. There is no evidence that selection caused the changes in gene frequency, nor is there evidence that it did not.

I established a permanent sampling schedule. The results are recorded in Table III.1.

TABLE III.1 Population Size and Frequency of White-Wing Allele

Site	Year 1			Year 2		
	N	p	$|\Delta p|$	N	p	$|\Delta p|$
1	102	0.41	0.04	634	0.43	0.02
2	873	0.64	0.02	>5000	0.66	0.02

In both populations we see a change in gene frequency each year. Could these changes be due to selection, mutation, migration, or drift?

Again, let us answer the question in checklist fashion. Remember throughout that we are only answering the question whether or not each individual force *alone* is a reasonable cause of the observed change.

As we learned in Chapter I, selection is always a possible explanation of changes in gene frequency, especially when the environment can change and allow for any possible number of fitness sets. Therefore, in our data, the change in frequencies may be due to selection, even in population 1 where the change differs in direction from year to year.

The change due to unidirectional mutation is

$$\Delta p = \mu p$$

Even when we look at the smallest change and solve for μ,

$$\Delta p = 0.02 = \mu(0.64)$$

$$\mu = 0.03$$

the mutation rate obtained is several orders of magnitude above the observed mutation rates. Therefore, we can say that mutation alone is not a probable cause of the observed changes in gene frequency.

We previously rejected migration as a force maintaining the polymorphism in the original population because of the observed natural history of the butterflies. Because we have no reason to believe that the natural history has changed, and since there were no strong winds during the sampling period, we can continue to reject migration as a possible force.

To test whether or not we can reject genetic drift as the solitary force molding gene frequency, we must carry out the same test, as earlier, comparing observed Δp's to the limiting values of $\pm 2\sigma$. To save time, the calculation of 2σ for the first year is shown in population 1 only. The other calculations are included in Table III.2.

To calculate the value 2σ for the first year of population 1, we must use the gene frequencies of the population the previous year, that is, the population that was sampled by reproduction. The population size N of the present year is the number of samples taken. Thus

$$2\sigma = 2\left(\sqrt{\frac{(0.45)(0.55)}{2(102)}}\right)$$

$$= 0.07$$

The rest of the values are determined in the same manner.

TABLE III.2 Absolute Values of Δp and Values of 2σ

Site	Year 1		Year 2	
	$\|\Delta p\|$	2σ	$\|\Delta p\|$	2σ
1	0.04	0.07	0.02	0.03
2	0.02	0.02	0.02	0.01

It is clear that the observed frequency changes in population 1 are within the range of $\pm 2\sigma$ for each year. Genetic drift may have been the cause of the observed changes here. Population 2, however, has a change in gene frequency in the second year that is greater than that expected by drift alone. Although this is not conclusive proof, it appears that another force, possibly selection, must be acting on this population.

Is it possible that selection is occurring in one population and not in another? Yes, certainly. Fitness is determined by the interaction of genotypes and environment. Different sites may render the same set of genotypes neutral with respect to each other or may result in selective differences. This is the difficulty for the experimental population geneticists, one with which they have learned to live.

Homework exercises

For one of the problems in the following set you will need a formula for the binomial distribution. If one trial has two outcomes with a probability p and $q = 1 - p$, then the chance of having in a sequence of independent trials the first outcome a times in n trials is

$$C_n^a p^a q^{n-a} = \frac{n!}{a!(n-a)!} p^a q^{n-a}$$

Of course, $p^a q^{n-a}$ is the probability of exactly a first outcomes and $(n-a)$ seconds in a specified order. However, there are C_n^a ways in which these a events can be distributed over n trials. This is the reason we have a multiple C_n^a in the formula.

The most important problems in this set relate to the issue of whether a particular change in allelic frequency can or cannot be explained by drift. To solve these problems, one should look at the allele frequency (not the genotypic frequency, do not mix them) and use our 95% reliability rule of $\pm 2\sigma$ to see whether or not the next generation frequency is within this range. The value of σ is given by the familiar expression

$$\sigma = \sqrt{\frac{pq}{2N}}$$

where p and q are allelic frequencies and N is the population size (number of individuals). If the allelic frequency is not given in the conditions of the problem, you can reconstruct it from the genotypic frequencies based on the Hardy-Weinberg law (see Homework Exercises, Chapter I).

Exercises 4 and 6 do not need any special reminders and assume that you read carefully the material of the preceding chapter.

1. You and I are playing a betting game in which a fair coin is tossed. If the coin comes up heads, I win a dollar; if it comes up tails, you win a dollar. After 200 tosses you have won $6 (103 of the coin tosses have come up tails). Because each of us had only $20 to begin with, I am becoming a little nervous about my finances. I decide to quit after 10 more tosses. What is the probability that I will win back my $6 by the end of the game? If we decide to play until the complete ruin of one of us, what are my present chances of eventually winning?

2. In a small population of 50 individuals the frequency of a certain allele is observed to increase from $p = 0.20$ to $p = 0.27$ in one generation.
   (a) Could this change be due to genetic drift? (Be 95% certain that your answer is correct.)
   (b) The same change is observed after *two* generations in a population of 200 individuals. Could this change be due to drift (with a 95% level of certainty)?

3. In a small town in Ohio (pop. 500) the proportion of blue-eyed (bb) people was 0.36. After two generations, the frequency was 0.49. The town was known to have families of only two children each on the average. After recognizing the rapid change in frequencies, several brown-eyed townspeople ($B-$) began to suggest that the blue-eyed people might have been enjoying more than their share of infidelities. The blue-eyed people said that this was libelous and that their increase was due to "the luck of the draw." Assume nonoverlapping generations and test the brown-eyed accusation (being 95% sure to avoid a libel suit).

4. A large population of fruit flies inhabits a coffee can filled with rotting fruit. A storm inundates the can with water and a large number of flies are drowned, but the population quickly rebounds to a high level when the water evaporates. If the population size were 2000 before the storm, dropped to 100, and then rose to 1000, what is the value of F (inbreeding coefficient due to drift) before and after the storm?

5. Desert pupfish is a species which inhabits hot springs in Nevada and other parts of the southwestern United States. Suppose that a species of pupfish that is confined to a single pool near Las Vegas is threatened by the construction of a new casino. In an attempt to save the species 50 breeding members are transferred from the pool to a similar isolated pool in a remote part of the state. Forty of these individuals have a mottled color pattern, which is determined by a dominant allele at a single gene locus. After one generation, a population census reveals that we still have 50 individuals, 38 of which are mottled. Could this change in gene frequency have been due to genetic drift, mutation, or natural selection? What if we had transplanted 500 fish instead of 50 and observed the same magnitude of change in gene frequency. Is the answer different from the first case?

6. According to the neutral theory of evolution, how many new mutations will be fixed over a 1-million-year time span in a species with a constant mutation rate, $\mu = 10^{-6}$? (Assume one generation per year.) In a population of 10,000 breeding members what is the level of heterozygosity predicted by neutral theory?

lecture 12
Today's view of evolution

A single lecture or, for that matter, a single lecturer could never truly give a catholic view of evolution today. By nature, the lecture must be to some extent parochial and the title should be "My View of Evolution Today," "Where I Stand and the Others," or "A Few Problems that Are Intriguing to Me and Others." I hope this will be more of the third and less of the first.

Specifically, I would like to touch briefly on three topics. The first is Sewall Wright's adaptive landscape and his general view of evolution of subdivided populations. While historically long preceding the present-day neutralist-selectionist controversy, it thematically dovetails nicely into our discussions. The second topic will touch on levels of selection. Actually a widespread topic, it has repeatedly cropped up as a fascinating problem over the years and will surely continue to do so. The third subject leaps into the arena of macro-evolutionary controversy and deals with gradual versus punctuated evolution. This will help to tie together some of the general concepts of microevolution into the grander process.

In Lecture 4 I spoke of R. A. Fisher, one of the greatest population geneticists of this century. Sewall Wright, a contemporary of Fisher, is a scientist of equal standing and influence. Wright, still active today, is presently Professor Emeritus at the University of Wisconsin–Madison. While it is almost absurd to decide which is Wright's most important contribution to evolutionary thought, two of his concepts that are well established are the adaptive landscape and the shifting balance theory.

In Lecture 5 I demonstrated the way in which fitness curves could be drawn for one locus, two alleles. This curve formed a ceiling under which our

Fundamental Theorem balloon wafted, seeking the highest point. Let us expand this concept to the entire genome of an organism. The average fitness of a population will change with the changes of the frequencies of many alleles. Instead of one independent axis in our graph, we have many. The average fitness functions describe not a line but a surface. Furthermore, consider the real possibility that genes interact so that the fitness of an allele can be radically changed by the composition of the rest of the genome. In such a case a population could have many fitness maxima over the range of possible genotypic compositions. The surface now becomes an *adaptive landscape* with peaks of high fitnesses and valleys of low fitnesses.

Conceptually, this is not much different from the idea of a fitness ceiling. The Fundamental Theorem restricts movements only up slopes and never down, so that selection drives populations to local maxima on the landscapes. Changes in environment have the same effect as in our previous model, causing "tectonic" shifting of our surface so that valleys become mountains and ridges become plateaus. The genetic composition of populations will follow this fitness landscape driven by natural selection. The genetic composition will move along ridges, skirt valleys and craters, and climb fitness peaks and mountains.

The unique point to this model and the shifting balance theory is that populations or species will not necessarily stay at a particular pinnacle of fitness and be isolated from all other adaptive peaks. Wright was the first to point out convincingly that fluctuations in population size and the resultant drift can "blow" a population from one peak onto the area of attraction of another adaptive peak. Small, isolated populations that are suddenly reduced in numbers by natural catastrophes may have significantly different genetic compositions from large, central populations. These differences occur because the ecological events of dispersal, isolation, or natural disasters all constitute genetic sampling events with small sample sizes. If the drift caused by these sampling events is large enough, a population may move away from a stable equilibrium peak. Drift will be counteracting selection. Then, as the population recovers and grows, drift will become less important and selection will take over as the dominant force. If the population has moved on the genetic landscape to the slope of a new fitness peak, selection will rapidly drive the population to a new adaptive equilibrium. This will cause shifts in the genetic makeup of the population or species over time. As a corollary, a species that is composed of small, semiisolated populations will be the most susceptible to evolutionary change. This concept of dynamic evolution, even when given fixed fitness structures, perhaps provides a golden mean to our evolutionary models.

In Lectures 2 and 3 the idea was presented that we are building models dealing with changes of gene frequencies of alleles in individuals. Although in some instances it is possible to disengage the allele from the organism, when we developed our models of selection, fitness was a characteristic of the genotype of an organism. In simpler words *the individual is the unit of selection.* This

Lecture 12   Today's View of Evolution   111

type of selection or evolution is generally discussed under the banner of individual or Darwinian evolution.

Is the individual the only possible level of selection? No, why should it be necessarily? Selection can conceivably exist on the levels both above and below that of the individual. For example, there is evidence of selection for DNA fragments, encoding genes, chromosomes, and gametes. These selection scenarios entail the disproportionate propagation of all these units independent of the propagation of the individual containing them. In other words these units can have phenotypes that are totally neutral on the level of the individual in which they exist. On the other hand, selection may possibly act on levels above the individual such as families, populations, species, or phyletic lineages. Genetically inherited properties may exist that differentially perpetuate such units while being neutral or even deleterious to the individual containing them. Of all these, I will key in on two such levels, one above and one below the level of the individual, and leave the rest to your own research.

The first level, *DNA sequences*, has received much attention recently as a result of new findings from molecular genetics. Within the genome of higher organisms there are long expanses of DNA that are composed of repeating, nontranscribed nucleotide sequences. Whether these are the result of selective reproduction of these stretches within the lineage or of repeated unequal crossing over is not clear.

Two other phenomena also fit within this level of selection. Transposable genetic elements, or *transposons*, are segments of DNA that independently insert and excise themselves within a host genome. They do not necessarily code for a gene product essential to the maintenance of the host cell, although they may carry genes that are expressed in the host phenotype or they may disrupt an essential function by inserting themselves within the active sequence of a host gene. Similarly, in the segments in the nucleolus organizer in primates there are repeated copies of the rRNA coding genes found on independent chromosomes. There is recent evidence that mutant alleles, logically, originating in only one site, have spread to the redundant sites throughout the genome in different lineages.

Although there is presently no clear relationship among these three phenomena, it seems as if there may be mechanisms where a DNA fragment may reproduce and increase in frequency totally independent of the host individual and its phenotype. It is winning the existential game totally and solely for its own sake. It is almost as if parts of DNA molecules can parasitize the genome of living organisms. This exciting area is only beginning to open up. Who knows, we may be seeing the start of neo-neutralism.

Familial or *kin selection* as it is better known is a step above individual selection. In this level of selection the group of related individuals is the unit of selection and the fitness of an allele is determined by the relative survival and growth of this group and not of the individual. The allele could even be deleterious to the individual and yet spread as in some models concerning altruism. *Altruism*, in the genetic sense, is a genetically controlled behavior that reduces the individual's fitness while increasing the genetic fitness of others. Using this definition, a rich philanthropist is not often altruistic, but a poor nun almost always is. Of course, this would assume that being a philanthropist or a nun is genetically controlled, which is not true. However, with a little good will and imagination, you can understand the genetic concept. In such a case the fitness of the allele is not determined by the fate of one individual, but by the cumulative fate of all individuals maintaining that allele.

For example, a herd of deer is composed of prereproductive juveniles, reproductive males and females, and postreproductive adults. I will make the following assumption concerning the social structure of the herd. Males develop a hierarchy early on so that the dominant male among the reproductive class mates with a greater proportion of the females than any other male and, on reaching the last year of reproduction, becomes the chief defender of the herd. Defense of the herd consists of the dominant male standing his ground against predators while the rest of the herd escapes. He is behaving like an altruist.

If this behavior pattern is determined genetically, what is happening to the genetic composition of the herd? First, in our example this defense behavior is to the great disadvantage of the defending male. He is running a greater risk of getting killed by defending the herd than by escaping with the rest. What is happening to the fitness of the male? He is lowering his fitness somewhat by possibly losing the chances of future reproduction, by getting killed. However, this behavior does increase the probability of survival of the rest of the herd. If the dominant male has produced a greater number of offspring than the other males, and the action of this gene indirectly increases the viability of his offspring, the gene may spread. This depends on the size of the herd and the proportion of the population that is related to the dominant male, a measure that is similar to our inbreeding coefficient. If the herd is large, most of the individuals benefiting (and thus increasing their individual fitness), do not share the dominant male's genes. The cost with which the individual is encumbered is greater than the benefit he receives by differentially promoting his offspring relative to others. This trait will not spread. However, if the herd is small and most of the deer do share his genes, then the phenotype of the gene (the behavior) stimulates the perpetuation and spread of copies of itself (fitness) in the population. This is analogous to our previous model of selection, except that it is the family unit rather than the individual whose fitness is being increased. Altruism works, then, if you "keep it in the family."

Units of selection that are higher than the familial level appear to be more problematic. Most models for group or population selection have logical weaknesses. The same phenomenon that the model intends to explain may often be better explained by individual selection. Without trying to be enigmatic, I suggest that you read up on the subject in a general evolution textbook. To conclude, let us discuss parenthetically a still higher level of selection:

species selection. I say parenthetically because the actual subject is the controversy surrounding the theory of punctuated macroevolution. Species selection is simply a corollary of this theory.

Macroevolution is the general term for gross changes in phenotype, almost exclusively morphological changes, over geologic time. This is probably your former conception of evolution. The generally accepted viewpoint of macroevolution, as stated by Darwin and as developed in the 1940s and 1950s by geneticists, zoologists, botanists, and paleontologists such as Theodosius Dobzhansky, Ernst Mayr, C. Ledyard Stebbins, and George Gaylord Simpson, was that of *gradualism*. Populations or species evolved by a gradual accumulation of new traits. The accumulation process was thought to be the result of all the various microevolutionary forces with which you are familiar from the three previous chapters. The fossil record is then generally expected to display even transformations from one type of species to another.

The fossil record does not always display such even transformations. Sudden changes in morphologies with gaps or missing links often exist. Within the gradualist framework, these discontinuities are considered to be the result of (1) an incomplete sample of the fossils themselves or, more substantially, (2) the result of sudden changes in selection pressures which cause relatively rapid genetic evolution. The changes are seen as being instantaneous when looked at in geologic time (such as 50,000 years).

The *punctuational model*, as developed by Niles Eldredge, Steven J. Gould, and Steven M. Stanley in the early 1970s, views these gaps not as sampling errors or special cases, but as the essential phenomena of macroevolution. Evolutionary change is expected to be rapid and concentrated, occurring only at a speciation event. During all other times, the species is expected not to change much. The fossil record then should show long stretches of uniformity or *stasis* (millions of years) interrupted or "punctuated" by sudden changes.

The two theories are totally compatible at this point since gradualists long ago pointed out that even, gradual evolution does not mean constant evolution. The rates of evolution are expected to change with changes in selection pressures, population size, and migration for reasons that you all know by now. The difference between the two theories comes in the mechanism behind the change and the factors leading to grand evolutionary trends in lineages of organisms.

The punctualists view speciation as occurring strictly in isolated populations located at the periphery of the species range. New, specific changes in characters rapidly occur. This then produces the jumps that we see. Each isolated population that arises is expected to evolve these jumps independently, so that the characters that do change are random between populations. To put this simply, one population may be distinguished by rapid evolution of furry ears and small teeth, while another isolated population coming from the same mother species may develop long legs and a pointed nose.

Yet general trends do occur in lineages such as the increase in leg length, reduction in toe number, and the lowering of the muzzle in horses. Putting aside the question of whether or not there are links missing in the lineage, which is not an essential argument, how do the two theories reconcile these trends? The gradualists would obviously say that the selective pressures for these cursorial adaptations remain consistent, though maybe not exactly constant, over geologic time. In other words natural selection on individuals produced a directional evolution. The punctualists would argue that actual character change is independent of such trends in terms of selection on individuals, but that, when new species interact, some are selected at the expense of others. It is this *species selection* that produces the trends, not individual selection. Microevolution is then decoupled from macroevolution.

Why talk about this controversy rather than about other subjects in evolutionary theory? First, this is a new "hot" topic. Yet I am obviously not unbiased and strongly believe that the controversy will be resolved shortly. Second, it does introduce the idea of selection above the family level, as mentioned earlier. Third, it does discuss the body of theory that I hope you have mastered in a grander or macro arena. Most importantly, it reintroduces the subject of scientific dichotomies that we discussed in Chapter II. Using this as our ultimate question, let us continue.

Is there a real dichotomy between gradualist and punctualist theories and which one do the data support? In some ways there are real differences. Punctualists would predict that the degree of character change or genetic change between species would be directly correlated with the number of speciation events that occurred between these groups, since genetic change must occur and will only occur with speciation. The paleontological data tend to be problematic for answering this question. Paleontological species are usually defined by morphological traits. Those traits that vary continuously are not good defining traits for species. Those that are constant and then abruptly change are good identifying traits. These traits thus are taxonomical markers for defining species. This is tautological since we define a species by the differences in these traits. So, of course, traits change with speciation. Modern species do not show a clear pattern of correlation of genetic and morphological change with speciation. Reproductively isolated populations may or may not display morphological or genetic differentiation.

What about the other predictions of the theories? Punctuated evolution depends on isolated, peripheral populations. Such a scenario can also fall within the realm of the gradualist theory. More importantly, other scenarios of speciation do exist and are theoretically possible. This would be inconsistent with punctualism.

Species selection is also a problematic concept. Any level of selection must assume a large *turnover* of groups at that level. This may sound more complicated than it is. If you assume individual selection, then there must be many individuals so that one type may replace the other without the population going extinct or being more affected by drift than by selection. Individuals can be and are produced and later die at fairly high rates. This is what I mean by turnover. Are species really produced and driven extinct so that there is a large turnover and thus selection between them? Are there traits that increase speciation and reduce chances for extinction on the level of the species? It is neither clear that there are such traits nor that there is a large turnover.

Does macroevolutionary theory divide into such neat dichotomies? In the end probably not. Punctuated evolution as a concept is important in reawakening an awareness of the variable rates and patterns of evolution. These obviously are not inconsistent with existing theory. The punctualist theory is fruitful in generating thought concerning levels of selection and evolution, though its proposals may need serious revision. The final shot has yet to be fired, and the victors are certain to be the scientific historians who study the development of science itself. Without doubt, its goal as *eparter le bourgeoisie* has been reached.

… # chapter IV

Population Growth

lecture 13
Malthusian growth

Before studying population growth models, let us discuss the relationship between these ecological models and the evolutionary models that preceded them. Your first clue to this relationship is the absence of a major division in these lectures declaring "Population Genetics" and "Population Ecology." Ecological population and genetic population models are commonly treated separately, often as two different disciplines. There is some historical element to this division since the models were developed independently, often by different workers. Beyond this, the separation is artificial. Ecological adaptations to the environment affect fitness coefficients and are, in turn, affected by selection. Similarly, the average fitness of a population $\bar{W}$ is the same as the average number of offspring per parent, a measure of population growth. On a different level, more involved models of selection, such as frequency-dependent selection and density-dependent selection (see Lecture 22), are essentially ecological-genetical models. Predator-prey or mimicry systems are not necessarily static from an evolutionary point of view and models simulating the evolution of such systems are a mixture of ecology and genetics. For heuristic reasons alone, these lectures are somewhat categorized. Where possible I will try to give a taste of the interface of the subjects; Lecture 22 is devoted solely to ecological-genetical models.

We will forget for awhile that populations are not homogeneous, that is, they consist of individuals with different genotypes, different ages, sexes, and so on. Let us think crudely of a population as a collection of individuals that is distinguished from any other collection (such as coins and stamps) by the basic property of reproduction, that is, the ability to make copies of themselves. They clearly need some material to make these copies so that our biological population is not a closed system. They use substances from the environment, such as food and oxygen, and transform them into "their own kind." We would like to count the number of copies they can make and to describe the growth of such simple abstract populations where every individual is exactly like the other one.

Let us assume again nonoverlapping generations, and call R the average number of offspring a parent can produce in its lifetime (one generation). In the case of a sexual population we will only be concerned with the females and assume that there are enough males to fertilize all of them. Female abundance in this case will be the limiting factor of growth and we will have a good enough picture if we count only mothers and their daughters.

Let us start with a population of N_o individuals. If everyone leaves on the average R_1 offspring in the first generation, we will have

$$N_1 = R_1 N_o$$

individuale after one generation. Conditions during the second generation might be different, for instance, it may be warmer or there may be less food available. Thus in the second generation the average number of offspring per parent will be R_2, with $R_2 \neq R_1$ in general. We will then have

$$N_2 = R_2 N_1 = R_2 R_1 N_o$$

We can continue to account for this basically multiplicative process over t generations to obtain

$$N_t = R_t R_{t-1} \ldots R_1 N_o$$

Be sure you understand that every R_i ($i = 1, \ldots, t$) describes both fertility and survival, so that the value of each R_1 gives the net result—the average number of offspring per parent that survives until the next reproductive period. We could have written every R_i as

$$R_i = B_i D_i, i = 1, \ldots, t$$

where B_i is the fertility of an average individual ($B_i \geq 0$) and D_i is the probability of survival ($0 \leq D_i \leq 1$) of the average individual. We do not want to go into such detail now, because we are interested in the population dynamics on such a gross level that the net reproductive rate is all that matters. You will see in a few lectures that more detailed models will take into account fertilities

Lecture 13 Malthusian Growth

and survivorships separately, but, for the moment, let us look only at the big picture.

Assume that the environment is constant and all R_t are equal, $R_t = R$. Our multiplicative formula simplifies in this case to become

$$N_t = R^t N_o$$

This is a discrete form of what we call the Malthusian law. Clearly, when $R > 1$, the population is growing; when $R < 1$, it is decreasing, and when $R = 1$, it is staying constant. The law says that in ideal, invariable conditions populations grow as geometric series with the multiplier R, which varies with the particular biological species and the particular environment. For example, if such a population of cells as *E. coli* reproduces by division where every cell makes two daughter cells in one generation, we will have, beginning with one cell ($N_o = 1$),

$$N_{10} = 2^{10} \times 1 = 1024$$

cells after 10 generations. This is an extremely fast growth. If this process continued, we would expect that the earth would be entirely covered by these cells in a short time. Indeed, the reason that the earth is not covered with *E. coli* or any other organism is that, of course, R does not stay constant for a long enough time.

The human population is probably the one exception in that R has stayed approximately constant for a long time. If we measure time in years, the value of R for the human population is about 1.03. The time it takes for the population to double in size can be determined by solving for t in the equation

$$2N_o = R^t N_o$$

$$t = \frac{\log 2}{\log R}$$

For our human population, the doubling time is about 23 years. This is also an extremely fast growth. If you look at the curve of the human population growth and try to see the effect of World War II, you can notice only a small bump on the curve. The 40 million killed during the War were replaced in no time. To take another example, before the potato blight of the 1840s there were approximately 8 million people in Ireland. After the blight struck, about a third of the population starved to death and another third emigrated, leaving about 2.5 million. Assuming that R was the present-day 1.03, it would have taken little more than 39 years to return to the preblight population size.

Geometric sequences grow quickly. One with which you are familiar is inflation of money. With an average inflation rate of 9% per year, money inflates twice in about 8 years, which means four times in 16 years and eight times in 32 years. We should all be millionaires very soon!

Thomas Malthus became famous, not for noticing this geometric series as a descriptor of population growth, but for indicating the fundamental consequences of such rapid growth for human society. This is similar to the Hardy-Weinberg law in population genetics. If it had not been Malthus, somebody would have noticed it. Why then do we call it a law? Did you notice that, until now, the title of law was only given to the simple Hardy-Weinberg formula? Now we apply it to the elementary formula of a geometric series. Both are extremely simple. Moreover, populations do not grow exponentially for any reasonable length of time. Why then is the Malthusian law so important?

Consider the first of Newton's laws, which says that in the absence of forces a body should move uniformly in a straight line. Do we ever have situations without forces? No, there are forces present all the time. Can we then verify Newton's first law experimentally? No, this is impossible. The value of this law is in setting the stage. It describes what would happen if nothing happened; it provides the background against which the other laws of physics take place. The Malthusian law has a similar role. It describes how populations would grow if environments were ideally constant, that is, the population would have no effect on the quality of the environment. The geometric series is the background for all other models of population growth. It stresses the multiplicative character of the growth process which is the most important feature of all reproducing populations.

Another mathematical form of the Malthusian law is more often used in models of population growth—the *continuous time form*. Let us consider population size N as a function of continuous time t, and introduce the instantaneous rate of change of the population size dN/dt, which is a derivative of the function N_t. For our purposes, it is enough to understand that dN/dt is the momentary rate of change of the function N_t at time t. The new form of the law is the following equation:

$$\frac{dN_t}{dt} = rN$$

This is the simplest linear differential equation describing the growth rate of a population as proportional to the population size at every particular moment. The parameter r describes the relative rate of growth:

$$r = \frac{1}{N_t}\frac{dN}{dt}$$

This is the rate of growth per individual in a population. As in the discrete model, r can be represented as the difference between the birth rate and the death rate, $r = b - d$. Thus the equation can be rewritten:

$$\frac{dN}{dt} = bN - dN$$

We are again interested only in the net result and, therefore, use r as an expression of the net combined effect of births and deaths. Both of them are assumed to be proportional to the population size.

Given the initial value for the population size $N_o = N$, we can solve the equation as follows:

Lecture 13   Malthusian Growth   121

$$\frac{1}{N_t}\frac{dN}{dt} = r$$

$$\frac{d}{dt}\ln N_t = r$$

$$\ln N_t - \ln N_o = rt$$

$$\ln \frac{N_t}{N_o} = rt$$

$$\frac{N_t}{N_o} = e^{rt}$$

$$N_t = N_o e^{rt}$$

The solution of the equation is an exponential curve as shown in Figure 13.1.

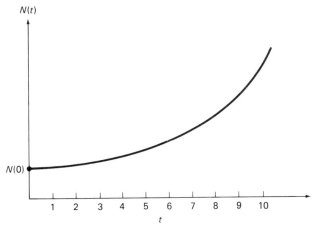

Figure 13.1 Exponential growth curve showing population size plotted against time.

The population grows exponentially when r is positive (birth rate exceeds death rate), decreases exponentially when r is negative (death rate exceeds the birth rate), and stays constant when $r = 0$ (birth and death rate exactly balance each other). The first case is the most unrealistic. Populations cannot grow indefinitely; something should happen to change the birth and death rates when populations grow large. These parameters cannot stay constant. We will deal with the models that account for the leveling off of the population size in Lecture 14. For the moment, to complete the picture of the Malthusian law, let us make a comparison between the two different forms that the Malthusian law takes: discrete and continuous.

Let us call our unit of time, τ. τ may be a week, a year, or 20 years, whatever is most convenient and sensible for the species that we are studying. In the discrete model we know that if we take the population size at a given time k, and compare it to the size at the next unit of time $k + \tau$, we will have

$$N_{(k+\tau)} = RN_k$$

or

$$R = \frac{N_{(k+\tau)}}{N_k}$$

Let us compare this ratio to our continuous model. Both population sizes can be exactly described as

$$N_{(k+\tau)} = N_o e^{r(k+\tau)}$$

and

$$N_k = N_o e^{rk}$$

R, the ratio of these, is simply

$$R = \frac{N_o e^{r(k+\tau)}}{N_o e^{rk}}$$

$$R = e^{r(k+\tau) - rK}$$

$$R = e^{r\tau}$$

Both values, r and R, are called Malthusian parameters and, depending on the context, one or the other is used to describe the population growth. It is important to remember the simple transformation rule which allows you to evaluate R given r and τ. The inverse rule can be obtained by taking the natural logarithm:

$$r = \frac{1}{\tau} \ln R$$

We will use this inverse transformation to evaluate r when given R and τ. Understand that the relationship between R and r strongly depends on the time-step of observation, τ, which is the subject of our choice. Depending on the question that we want to answer, we will always choose a convenient time-step to make the formulation easier. You should also notice from this transformation the relationship between fertility B and survivorship D of the discrete equation and the birth rate b and death rate d of the continuous equation. If

$$R = BD$$

and

$$r = \frac{1}{\tau} \ln R$$

then

$$r = \frac{1}{\tau}(\ln B + \ln D)$$

The survivorship parameter D is always less than one so that the natural log of it is negative. From our definition of r, $r = b - d$, we see that

$$b = \frac{1}{\tau} \ln B$$

and

$$d = \frac{1}{\tau} |\ln D|$$

Malthus, the author of this simple but fundamental law, did his work at the end of the eighteenth century, about 200 years ago. Although the formula itself is quite trivial and was known hundreds of years before Malthus (in the discrete form) and even applied to the reproductive process, Malthus was the first to appreciate the practical consequences of this explosive growth for the human population. He anticipated a number of overpopulation problems, many of which we are facing today. Even though it is less noticeable than, for example, the problem of preventing nuclear war, it may be as important for human society as a whole. The book, *The Population Bomb*, by Paul Ehrlich of Stanford University, has heightened awareness of the problem. A number of highly developed, industrial countries now show a decline in the birth rate and even a net negative growth ($r < 0$, $R < 1$). This is a hopeful sign that sociological mechanisms, whatever they are, will take care of the problem. At the moment, a number of mathematical models that are supposed to account for future human population growth, food and energy supplies, and so on, predict serious difficulties in the world in general at the beginning of the twenty-first century. One should not trust these models too much; unexpected discoveries may shift the parameters of these models to a sufficient degree so that the present forecasts might be invalidated. The problem exists, however, and the absolute number of hungry people is growing.

The mathematical complexity of a formula rarely correlates with its practical importance. Simple models often prove to be of much greater importance than complex ones. Malthus was the first to see significant meaning in the simple geometric series and it is only right that his name is attached to this fundamental regularity governing population growth.

lecture 14
Logistic growth

Lecture 13 developed what is probably the strongest, most explosive force in ecology: Malthusian or logarithmic growth. In its simplest form it says that the rate of increase of a population is directly proportional to the population size, or

$$\frac{dN}{dt} = rN$$

where r is the Malthusian parameter, or instantaneous rate of per capita growth. When we integrate this we get the exponential form

$$N_t = N_o e^{rt}$$

In the discrete form we rewrite this as

$$N_k = R^k N_o$$

If these formulae are so basic and powerful, then the obvious question would be, how often does the growth of natural populations follow this law? The answer is almost never. If every or any population of any organism would continue increasing in numbers as described by the Malthusian law, these individuals would soon blanket the earth. There would physically be no place for anything else. Furthermore, there would never be enough resources to support such growth. Besides energy, organisms need carbon, nitrogen, phosphorus, and metal ions, all of which have a finite limit. A population simply cannot increase indefinitely in such a manner. In fact, most populations, with the conspicuous exception of humans, maintain a value of r close to zero. Those that do exhibit exponential growth usually have recently invaded an area and are exploiting an untapped, and thus momentarily nonlimiting, set of resources.

Let us begin thinking about a more realistic model by looking at a hypothetical population that has recently invaded such an untouched area. A number of beetles, both male and female, are isolated on a broken log, floating down a river and out into the sea. After 2 months, this wretched refuse lands on a previously beetleless island. They start exploiting the island and, after restoring their health, begin reproducing. The fecundity is low at first but increases during the second and third generations. The rate of growth of the population then starts slowing down even as the population continues to increase, because of the decrease of available resources per individual. Eventually, the population stops growing and remains at a relatively constant size. I have plotted population number versus time and population growth rate versus time for these beetles in Figures 14.1 and 14.2, respectively.

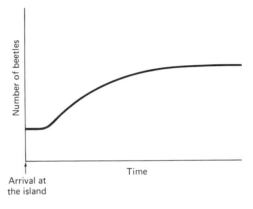

Figure 14.1 Size of hypothetical beetle population plotted against time.

Figure 14.2 Growth rate of hypothetical beetle population plotted against time.

Thinking in terms of our previous model, let us analyze why population growth, dN/dt, slows to zero. The rate of birth b, which was independent of the population size in the Malthusian model, would not be expected to be constant. Each beetle needs a certain amount of food to grow, to run around, and to carry on its general beetlelike behavior. In addition the beetle must have available another quantity of energy to reproduce. As the population increases, the resources become limiting, that is, there is not an overabundance of food, energy, or space per individual. Reproduction will go down. At the same time, the rate of death d increases because of the resource limitations.

When the birth and death rates exactly balance each other, dN/dt will be equal to zero. To generalize, with birth rate decreasing with population size and death rate increasing, the growth rate will be a decreasing function of population size. We can summarize by saying that birth, death, and growth rates are *density dependent*. By density I mean the number of individuals per square meter or some other appropriate unit of habitat size.

How can we describe this function mathematically? There is no a priori way of determining how birth and death rates change with population size, and there is no reason to expect that they will be described by simple functions. The dependence is probably different for different species and different habitats. We can only say that growth rate is some decreasing function of population size, or $dN/dt = N\psi(N)$, where $d\psi(N)/dN < 0$. Qualitatively, however, we can get a good understanding by describing the function as being *linear*. I am purposely stressing this point so that you realize that the model that we will develop, which is the model most often used, is linear for heuristic and technical reasons only. There are many reasons to believe that growth functions are often not linear in N. For example, many populations must have a minimum size to start growing. This phenomenon is known as the *Allee effect*. The clearest case is where there is cooperation in food gathering or juvenile care. If there are not enough individuals, the group may starve or the juveniles may die. Another example is when there are not enough individuals to find mates. If one male fly colonizes a new area, it is clear that this is not a sufficient beginning for a thriving and expanding population. In organisms in which population density plays a large role in reproductive success, such as in wind-pollinated plants, higher numbers than one may still be insufficient for population growth. With these caveats in mind, we still may learn a lot by developing our simple growth model.

I will now plot the growth rate as a linear function of population size, as shown in Figure 14.3. The simplest equation to describe this function is

$$\frac{dN}{Ndt} = r - \gamma N$$

or

$$\frac{dN}{dt} = rN - \gamma N^2$$

where N is the population size, r is the intrinsic rate of increase, and γ is the density factor. You can think of γ as the degree to which one individual takes away resources or reduces the growth of the population.

As a reminder, I have included the per capita growth function of Malthusian growth in Figure 14.3. The line is straight with no slope, that is, γ is zero; an individual has no effect on the level of resources available to the rest of the population. This growth is therefore *density independent*.

Let us return quickly to our new function. When population size is small, the per capita growth rate is close to r. As N increases, the growth rate decreases.

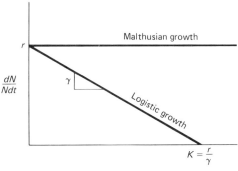

Figure 14.3 Per capita growth rate plotted against population size.

When we solve our equation, we see that dN/Ndt is zero when the population size N is equal to some number, r/γ. At this population size, the birth rate is exactly balanced by the death rate, and the population size reaches a constant value. This population size (r/γ) is commonly referred to as the *carrying capacity* and denoted by K. Using this notation, we can rewrite our growth equation to what is a more common form in the literature.

$$\frac{dN}{dt} = (r - \gamma N)N$$

$$\frac{dN}{dt} = rN\left(1 - \frac{\gamma}{r}N\right)$$

$$\frac{dN}{dt} = rN\left(1 - \frac{N}{K}\right)$$

This equation is often referred to as the *logistic equation* and the growth function as *logistic growth*, although other names are used.

The carrying capacity of a population and its intrinsic rate of growth are powerful concepts and, because of that, are often misused. First, you should intuitively understand that there is no such thing as the carrying capacity *of a population*. Rather, it is the carrying capacity of the environment with respect to a particular population. Simply, if an environment is rich and mild for a given species or population, it will support more individuals than a depauperate and harsh environment. The carrying capacity of the first is greater than the carrying capacity of the second. Second, there is little that is intrinsic about r for a population or species. For example, the more nutrient rich the environment, the greater the fecundity of individuals. This factor is totally independent of any density effects. In other words the resources available may not be limiting to the individuals, but other important resources may not be present at all. Thus both r and K (or γ) may be influenced independently by the environment.

Of even greater importance is the relation between r and K. When quantitative field ecologists began collecting a large body of data on r and K values of natural populations, it became apparent that r and K were correlated. The species with a higher r tended to have higher K values, too. The reason that this seemed so surprising is actually historical. To begin, the logistic equation is most often written in terms of r and K. In this equation r and K look like two independent parameters. There seems to be no reason to think that they are correlated. However, if we use the equation containing our density factor γ, it is clear that the carrying capacity is directly proportional to r.

$$K = \frac{r}{\gamma}$$

When r increases, K increases. Second, there is an indirect relationship between r and K because the environment produces similar effects on them, as we discussed earlier.

A more important reason for the surprise was the development and influence of the concept of r and K selection. In the early 1960s Robert MacArthur suggested that species or populations that exploit rapidly changing environments and also experience density-independent mortality should evolve to increase their intrinsic rate of increase r. This makes intuitive sense if you think of a weed growing in a plowed field. It must establish itself, produce a lot of seeds, and send those seeds to another habitable environment before it is plowed under by Farmer Jones. The problem of how well it can grow and reproduce under crowded conditions, that is, density-dependent growth, is irrelevant since the population rarely sees such situations.

In contrast to such a strategy, a population that grows in a stable environment and reaches greater densities should evolve to reduce the effects of other individuals competing with it. Using our notation, the γ should decrease, allowing for a higher K or carrying capacity.

A simple example of this would be grasses, such as wheat or barley, that have leaves perpendicular to the stem and parallel to the ground. When growing in the wild, such a structure would be effective in capturing light but it would also be effective in shading a neighbor. The plants would be competing for light and only a certain density could be reached. By selecting for leaves that are drooping, it is possible to reduce the shade canopy of each plant. This reduction in shading ability reduces competition, that is, reduces γ. If r remains constant, the carrying capacity will increase and the farmer can sow denser fields.

As a basic concept for looking at evolutionary ecological strategies of plants and animals, r and K selection is helpful. Unfortunately, the concept has often been used as a method of classification, rather than as a method of comparison. Some species are considered r strategists while others are K strategists. There is a fine difference here; there is no obvious trade-off between r and K. Indeed, as mentioned before, if γ is constant, K will increase with an increase in r. Clearly, depending on how r and γ change, K may remain constant, decrease, or increase. Therefore, a species can evolve in the direction of increasing both r and K simultaneously.

Let us explore the logistic growth equation by using it in a semi-realistic

Lecture 14   Logistic Growth

situation. Consider the problem of how to harvest a population optimally. Because of the simplifications built into the logistic growth model, this is not going to make you great fishermen instantly. On the other hand, you should leave this lecture knowing how to avoid being bad fishermen.

You have a population of fish that follows a logistic growth model. You want to catch the largest possible number of fish over time. How do you do this? If you catch all the fish in 1 year, there will be no fish for the next year. Similarly, if you take more fish than are being naturally added to the population each year, you will drive the population extinct. If you take less than are being replaced by reproduction each year, you will neither be making an impact on the population nor will you be taking a lot of fish. I want to take the same number that are being replaced *and* I want to do this at the level of highest replacement rate, that is, when dN/dt is at its maximum.

From the first semester of calculus you should be familiar with the procedure of getting our solution. We should find the population number that maximizes population growth by taking the derivative of dN/dt with respect to N and solving the algebraic equation by setting this derivative to zero.

$$\frac{d\left(\frac{dN}{dt}\right)}{dN} = \frac{d\left[rN\left(1 - \frac{N}{k}\right)\right]}{dN} = 0$$

$$0 = r - 2\frac{Nr}{K}$$

Solving for N, I get

$$\frac{2Nr}{K} = r$$

$$N = \frac{K}{2}$$

The second derivative

$$\frac{d^2}{dN^2}\left(\frac{dN}{dt}\right) = -\frac{2r}{K}$$

is negative, so $N = K/2$ represents the population size giving the maximum growth. The actual growth and thus my actual harvest each year is determined by evaluating dN/dt where $N = K/2$.

$$\frac{dN}{dt}_{\substack{\text{optimal}\\\text{havesting}}} = rN\left(1 - \frac{N}{K}\right)$$

$$= r\frac{K}{2}\left(1 - \frac{1}{2}\right)$$

$$= r\frac{K}{2}\left(\frac{1}{2}\right)$$

$$= \frac{rK}{4}$$

If I harvest more than this amount, I will drive the fish population to extinction. If I harvest less, I will not be catching as much as I could. You should realize, of course, that this simple rule of half the carrying capacity is based strictly on the logistic equation. With more complex, nonlinear shapes of density dependence, the optimal point will exist somewhere below K but it may be $(1/3)K$ or $(3/4)K$ or any value depending on the particular density-dependent function. There are many generalizations of the logistic equation. Most are more complicated, nonlinear ways in which population density influences its own growth. Mathematically, we can express this as follows:

$$\frac{dN}{dt} = N\psi(N)$$

where $\psi(N)$ does not have to be $(r - \gamma N)$, as in a logistic equation, but can be a more complex function. If we assume that $\psi(0) = r$ and that ψ decreases with increasing N, that is,

$$\frac{d\psi}{dN} < 0$$

(two basic properties of our logistic model), the qualitative picture of population dynamics stays basically the same. Figures 14.1, 14.2, and 14.3 will look qualitatively similar except that in Figure 14.3 the decrease of the rate of growth toward K will not follow a straight line. It will be a curved but decreasing function as well.

The qualitative difference can appear if we take into account the previously mentioned Allee effect. The essence of the effect is that, for the range of low population densities, an increase may be beneficial for the population growth rate, rather than detrimental. For very low population densities, the overall growth rate might even be negative. It would mean that populations starting at a very low density will not survive at all.

The qualitatively different picture for this case is given in Figure 14.4.

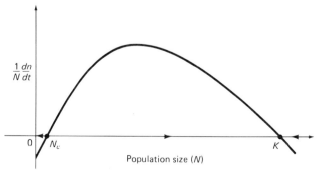

Figure 14.4 A nonlinear population growth function, demonstrating the Allee effect. Per capita growth rate is plotted against population size.

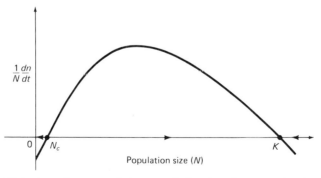

Here, the function $\psi(N)$ is not only nonlinear, but it also starts out negative, becomes positive only when $N > N_c$, and then decreases when N tends to the carrying capacity value. The arrows in Figure 14.4 indicate that if a population is initiated below the critical value N_c, it goes extinct. Therefore, $N^* = 0$ becomes the second stable equilibrium in addition to $N^* = K$. The value of $N^* = N_c$ is the unstable equilibrium. Population trajectories will be similar to the old case of the logistic growth only if the population is initiated above the critical level.

I have tried to make you circumspect, pedantic, and prudent (quite the golden mean). Two-population interactive models are a simple step away. These include predation, competition, and mutualism and are the subject of Chapter V.

Problem solving

Let us withdraw from the wilds of mountain tops or the thundering waves of a rocky shore to study basic growth equations. What is needed is an unexploited, pristine environment in which a population may figuratively stretch its wings. What better organism and more virgin site than a freshwater snail and a freshwater aquarium? Anyone who has tried to maintain an aquarium knows that the population growth of algae or snails can be explosive. On a small scale this is a classic example of a species invading a new area and finding unlimited resources, no competition, and no predation—the best of all possible worlds.

My aquarium snail population got its beginning from such a chance invasion. To increase the oxygen content of the water and, more importantly, to provide cover and environmental complexity for my fish, I introduced some aquatic vascular plants. Unknowingly, I also introduced six snails. After 2 weeks, I noticed 39 snails in the tank. Assume that half the snails are female. Based on these two data points, what is the rate of increase R, what is the intrinsic rate of increase r, and how many snails should I expect after 4, 6, 8, and 14 weeks?

The rate of increase R for the 2-week period is calculated by taking the ratio of the two population sizes. You can see this by rearranging the Malthusian growth formula.

$$N_1 = R^1 N_o$$

$$R^1 = \frac{N_1}{N_o}$$

R is the rate of increase and N_1 and N_o are female population size. Notice that the subscript 1 of population size and the power 1 of the rate of increase refer to one unit of time. We have arbitrarily designated 2 weeks as our unit of time. The rate of increase is

$$R_{2\text{ weeks}} = \left(\frac{39}{6}\right)\left(\frac{\frac{1}{2}}{\frac{1}{2}}\right)$$

$$= 6.5$$

The two factors of 1/2 are the expected proportion of females. Because we do not actually count the females but assume that they occur proportionately every generation, we may ignore this factor. The proportionality constant will factor out anyway, so we can work on total population size, not just the number of females.

We can determine the intrinsic rate of increase r by using the transformation formula

$$r = \frac{1}{\tau} \ln R$$

where τ is the time period, in our case 2 weeks.

$$r = \frac{1}{2} \ln 6.5$$

$$= \left(\frac{1}{2}\right)(1.8718)$$

$$= 0.9359 \text{ per week}$$

We can now determine the projected population size by two equivalent methods, the discrete or continuous model. For no particular reason, I will use the discrete description. Four, 6, 8, and 14 weeks correspond to two, three, four, and seven units of time, respectively. Therefore,

$$N_{4\text{ weeks}} = N_2 = (6.5)^2 N_o$$
$$= (6.5)^2 6$$
$$= 254$$
$$N_{6\text{ weeks}} = N_3 = (6.5)^3 (6)$$
$$= 1648$$
$$N_{8\text{ weeks}} = N_4 = (6.5)^4 (6)$$
$$= 10{,}710$$
$$N_{14\text{ weeks}} = N_7 = (6.5)^7 (6)$$
$$= 2{,}941{,}337$$

The thought of having almost 3 million snails after another 12 weeks was too much for me. I did the only reasonable thing possible: I left the laboratory. I asked a student to feed the fish, maintain the tank, and, as an aside, keep an eye on the number of snails in the aquarium. I fully expected to receive a frantic message from her that the snails were taking over the building or, worse yet, to get a letter from her parents suing me for the death of their daughter caused by the crushing by millions of snails. Fortunately, I received no such letters. After 18 weeks I returned to the laboratory.

The aquarium was crowded with snails, to be sure, but there were not the millions that I had expected. I sought out my student assistant and asked her what had happened. She said that the snails had increased in numbers rapidly during the first few weeks, but after the eighth week they had hardly increased at all. Table IV.1 lists the number of snails for every 2-week period up to 18 weeks.

TABLE IV.1 Populations Size of Aquarium Snails

	Week									
	0	2	4	6	8	10	12	14	16	18
N	6	38	231	995	2028	2414	2486	2498	2499	2498

Let us plot the population size as a function of time. What happened to the growth of the snail population? Estimate the appropriate parameters (r, K, γ).

I have plotted the population size as a function of time in Figure IV.1. The shape of the curve is sigmoid, common for logistic growth. The population reaches a plateau at approximately 2500 individuals and remains steady at that level for about 6 weeks. I have also calculated the per capita growth dN/Ndt for each population size by determining the ratio N_{t+1}/N_t and transforming this by means of the equation given in Lecture 13.

$$R_t = \frac{N_{t+1}}{N_t}$$

$$\frac{dN}{Ndt} = \frac{1}{t} \ln R$$

I have plotted out these values on Figure IV.2 as a function of population size.

It is clear that the growth of the snail population is not unlimited, but rather is affected by some density-dependent controls. At this point, I cannot determine what is limiting growth, but it is clear that the population is not exceeding 2500. Using this number as an estimate of the carrying capacity K of the population, I can determine the density-dependent coefficient γ.

$$\gamma = \frac{r}{K}$$

Chapter IV   Problem Solving

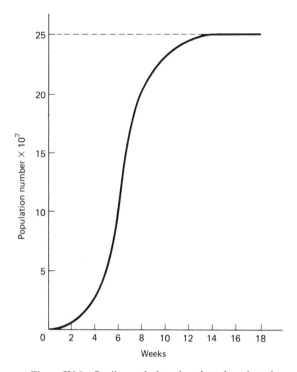

Figure IV.1   Snail population size plotted against time.

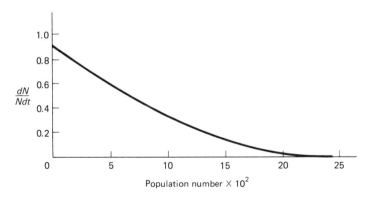

Figure IV.2   Per snail growth rate plotted against population size.

$$\gamma = \frac{0.9359}{2500}$$

$$\gamma = 3.7 \times 10^{-4}$$

I was pleased to find that my tank was not overflowing with snails, as a cornucopia molluska. Actually, I was so pleased that I decided to play around with the aquaria to get a hint of population mechanisms. Because the snails were almost exclusively found on the walls of the tank, I decided to see whether the volume of the tank or simply surface area affected population size.

The tank that I had used had the dimensions 30 centimeters × 50 centimeters × 25 centimeters. Thus the volume was 37.5 liters and the surface was 4500 square centimeters. I then constructed a set of tanks having the same surface area, with dimensions 25 centimeters × 45 centimeters × 45 centimeters, but the volume was now 50.6 liters or more than a 33% increase. I put together another set of tanks having the same dimensions as my first but with four 30 × 20 glass plates placed inside. The plates were placed so that fish and snails could freely reach all parts of the aquarium. In this manner I maintained the water volume of the original tank but increased the surface area to 9300, more than twice the original area.

I then introduced 100 snails into each tank and maintained as similar a regimen as possible. I again asked for the services of a student assistant and recorded the population sizes for the two treatments for 18 weeks. The results are listed in Table IV.2.

Without any probing analysis, it is clear that volume has no effect on the carrying capacity of the snails when treatment 1 is compared to our original aquarium population. Surface area does have a profound effect, however, as can be seen when treatment 2 is compared to our original population. It is especially of interest that the carrying capacity of treatment 2 is disproportionately larger than the simple increase in surface area. Surface area increased by a factor of 2.07 while carrying capacity increased by a factor of 2.4.

The surface area correlation immediately clarified my thinking. Snails are often introduced into an aquarium to keep the glass walls clear of algae. The snails scrape off the algae with their radula. The volume of the water has nothing to do with their food source. The only obvious factors that volume would control would be the concentration of oxygen and some minerals such as calcium. The water is saturated with oxygen from an aeration system. Because there is no obvious volume effect in the first treatment, I assume that calcium also is not limiting.

This still leaves the problem of why the carrying capacity of the snails in treatment 2 does not increase by the same factor by which surface area increases. Let us assume that the snails are harvesting the algae at maximum efficiency and that the algal growth is directly proportional to surface area. Is there a hidden theoretical factor that is disproportionately increasing my expected yield?

> Before I answer this question as I have posed it, I want to state clearly several assumptions. The first is that the reproductive capacity of the snails, as measured by r, has not been affected by any of our treatments. If I were to find the parameters for the differential equation for logistic growth given the trajectory of the snail population size over time,

TABLE IV.2 Snail Population Sizes for Two Surface/Volume Treatments

Treatment	Volume (liters)	Surface Area (cm²)	Population Size/Week									
			0	2	4	6	8	10	12	14	16	18
1	50.6	4500	100	533	1594	2299	2467	2499	2500	2500	2500	2500
2	37.5	9300	100	595	2503	4938	5807	5994	5997	5999	5999	5999

I would find that in this case r did not change. The change in carrying capacity comes from a change in γ, apparently because of a simple increase of food.

Given these assumptions, let me attack the problem in terms of the algae. If the snails are harvesting the algae so that they are maintaining a maximum crop, then, as we have shown in Lecture 14, the algae are maintained at a level of $K/2$ or the equivalent $r/2\gamma$. The actual maximum yield is $rK/4$ or $r^2/4\gamma$. Up to now we have been assuming, at least for the snails, that the carrying capacity K has been increased by a reduction in the density-limiting form γ of the aquarium and that this γ is directly proportional to surface area. What would happen if γ of the snails and the treatment is proportional to the maximum algal yield? Again, as a simple assumption, let us assume that K, the carrying capacity of the algae, only increased proportionally with the increased surface area. Are there any factors that the theory may point out to explain this paradox?

We know that K can increase either by an increase in r or by a decrease in γ. If we divide our original algal carrying capacity K_o by the algal carrying capacity of treatment 2, K_2, and set this equal to the ratio of surface areas, we get

$$\frac{K_o}{K_2} = \frac{\text{Area}_o}{\text{Area}_2} = \frac{4500}{9300} = 0.48$$

This, of course, is equivalent to

$$\frac{\frac{r_o}{\gamma_o}}{\frac{r_2}{\gamma_2}} = 0.48$$

Let us set the ratios of the snail carrying capacities of both treatments equal to the ratios of the optimal algal yield.

$$\frac{\frac{r_o^2}{4\gamma_o}}{\frac{r_2^2}{4\gamma_2}} = \frac{K_{o\,\text{snail}}}{K_{2\,\text{snail}}} = 0.42$$

$$\frac{\frac{r_o^2}{\gamma_o}}{\frac{r_2^2}{\gamma_2}} = 0.42$$

Substituting our previous ratio into this equation, we get

$$\frac{r_o\left(\frac{r_o}{\gamma_o}\right)}{r_2\left(\frac{r_2}{\gamma_2}\right)} = 0.42$$

$$\frac{r_o}{r_2}(0.48) = 0.42$$

$$\frac{r_o}{r_2} = 0.88$$

$$\frac{r_2}{r_o} = 1.14$$

Thus by increasing the intrinsic rate of growth by less than 15%, we can account for the disproportionate increase in the snail population.

As mentioned in Lecture 14, r, as well as γ, may be changed by improved environmental conditions. One possibility in our aquarium example is that the increased surface area of glass increased the concentration of available silicon in the water and allowed for an increased intrinsic growth rate of freshwater diatoms. Thus the increase in surface area, while only affecting algal carrying capacity linearly, increased the optimal algal yield quadratically.

My aquarium has led us into the realm of multispecies interactions. Some assumptions here may be too simple for real interactions, especially that of snails harvesting so efficiently. Usually, predators are not so efficient. In Chapter V we will examine extensions of our basic growth models to two interacting populations.

Homework exercises

One important point to remember when solving problems on Malthusian and logistic growth is that the values of both parameters r and R depend on the unit of time τ. The relationships are

$$R = e^{r\tau} \qquad r = \frac{1}{\tau} \ln R$$

Think always what is τ for every particular problem and then solve the problem in terms of R or r, whatever seems to fit better for the solution in the particular case.

The expression of the carrying capacity, K through r and γ,

$$K = \frac{r}{\gamma}$$

is all you need to solve the "static" problems related to logistic growth. I have never asked any dynamic questions in this case, although they can also be answered in a relatively simple way. Remember that, for the logistic equation, the level of the optimal sustainable yield (optimal harvesting) is given simply as

$$N_{opt} = \frac{K}{2} = \frac{1}{2}\frac{r}{\gamma}$$

1. If the human population of a country increased from 2 million to 4 million in 30 years, what is the Malthusian parameter r calculated on a per year basis?
2. A population consisting of 10,000 individuals has increased at a rate of $r = 0.5$ per year for the past 10 years. What was the initial population size? What is R (per year) for this population?

3. A population of guppies leveled off, after initial growth, at 60 individuals in the tank, and seemed to remain at that level for several months. If the birth rate is 14 offspring per parent per month and the death rate is 13 individuals per month, what is the value of the density factor γ? Suppose that I start rearing guppies in a larger tank and the population stabilizes at 100 individuals. If r has not changed, what is the new γ?

4. A population of butterflies has two generations per year. After assessing this population for several years, we calculate the following values for R (growth rate per generation):

Year	1		2		3	
Generation	1	2	3	4	5	6
R	1.5	1.2	0.8	0.5	2.5	0.2

In which years did the population increase? What is the average value for R (per generation) over the 3 years?

5. A colony of termites consisting of 10^3 individuals inhabits a rotting log in a deciduous forest. Assume that the colony exhibits logistic population growth and is now at its carrying capacity (K). Another termite colony living in a different log in a pine forest also grows logistically. This log is larger, however, and thus the parameter (γ) which describes the intensity of density dependence is one-half that of the deciduous forest colony. In addition the pine forest colony has evolved a rate of increase (r) which is 1.5 times as large as that of the deciduous forest colony. What is the carrying capacity (K) for the pine forest colony?

6. Suppose that colonies of ants that prey on termites are introduced in both areas. At what population size in each habitat should the ants maintain the termite colonies to obtain the optimal yield of prey? How many more termites per unit time can the ants in the pine forest take in comparison to the ants in the deciduous forest (at the optimal yield)?

7. The human population of the People's Republic of China is now estimated to exceed 1 billion. If the population is growing at a rate of about 2% per year, what is R? At this rate of growth, how long will it take for the population to double?

8. A population of *E. coli* increases at a rate of $R = 2$ per 20 minutes. After a period of 2 hours, the population size is 10^6. What was the initial population size? What is r (per minute)?

chapter V

Interacting Populations

lecture 15
Lotka-Volterra competition models: competitive exclusion

Two of the essential underlying, if unspoken, assumptions of the models developed in Chapter IV are the uniformity of all individuals in a population and the instantaneous response of the population to the environment (on the time scale accepted in the model). In the real world both of these assumptions are always incorrect to some extent. The degree to which the population in question deviates from these assumptions determines the degree to which the population growth curve deviates from the expected logistic equation or its analogue.

What are the common deviations from these assumptions? I hope that the first that comes to your mind is that individuals are usually genetically different. These genetic differences may result in absolute fitness differences. Let us briefly review what this means in terms of population growth. The absolute fitness W of a genotype is the number of viable offspring produced by that genotype. Suppose that we have a population of N individuals of which N_i is the number of individuals of genotype i in the population:

$$N = \sum_{i=1}^{n} N_i$$

Each genotype has its own absolute fitness W_i. The total number of offspring produced is then

$$N' = \sum_{i=1}^{n} W_i N_i$$

This, in turn, is equal to

$$N' = RN$$

Thus

$$R = \frac{1}{N} \sum_{i=1}^{n} W_i N_i = \sum_{i=1}^{n} W_i \frac{N_i}{N}$$

You should recognize that N_i/N is simply the frequency of the genotype in the population and, therefore, the right side of the equation is the weighted mean of the absolute population fitness. Put more succinctly,

$$R = \bar{W}$$

$\bar{W}$ can and often will change from one generation to the next. So, too, we should expect that density-dependent influences (γ in logistic equation) will change when under evolutionary pressures. If r or γ changes, the population will not follow logistic growth. A more detailed treatment of this interface between genetics and ecology will be given in Lecture 22.

Genetic differences aside, there are often developmental differences between individuals in a population that can lead to deviations from our simple growth models. For example, if the population is made up of reproductive individuals and prereproductive individuals, then the growth rate is affected at the same time by two different population sizes. The birth rate is affected by the number of reproductive individuals and the death rate by the total number. Instead of

$$\frac{dN}{dt} = rN = (b - d)N$$

we have

$$\frac{dN}{dt} = bN_{\text{reprod.}} - dN_{\text{total}}$$

This can become more complicated if the different life stages have different resource demands. For example, assume that each animal partitions the resource it uses into growth and development, maintenance and reproduction. As a consequence, the resource demands of individuals will be dependent on the life stage. Newborn individuals may require more resources to grow than adults to maintain themselves. In turn, reproducing adults may need more resources than nonreproducing adults, but perhaps less than the young. Thus our γ coefficient, which is a measure of how much an average individual inhibits the growth of the population, is totally dependent on the age structure of the population at a given time. For example, suppose that we have a population of reproducing adults. They are using the resources to reproduce but are still below the level at which all the resources are consumed. All reproduce at once, leave a single offspring, and then die. The population size has remained constant, yet the resource requirement per individual has changed. Because young individuals require more resources for development than adults require for reproduction, the resource utilization is above the maximum possible. Because the offspring cannot develop, they die and the population crashes. When there are overlapping generations, this scenario is less drastic, but even so, populations can *overshoot* or temporarily exceed their carrying capacities. The population responds by an increase in mortality until it is below carrying capacity. Eventually, the population begins to increase again and the process repeats itself, causing the population to *oscillate* above and below carrying capacity. Only when a population reaches a particular balance mixture of life stage classes will the response be balanced and the oscillations dampen out. General problems of age structure in population growth will be dealt with in Chapter VI.

This last example also touches on the problem of *time delay* in the response of the population to the environment. Many organisms have a feedback mechanism that acts to reduce reproduction when density increases. The greater the difference in resource utilization of the different life stages, the greater the time delay in the feedback mechanism and the greater the oscillations. Similar oscillations in population size can occur when the environment does not respond directly to the population's pressure. This is most often the case when the important part of the environment in question is another population of organisms, such as in a predator-prey system. Here population growth depends on both population sizes at a given time. At any given future time a population will be at a size determined by some condition at a previous time. This inherent delay between the response of growth and the actual population sizes can also lead to oscillations, especially when two independent growth equations are involved. We will study this case of population interactions in Lecture 16.

The purpose of this lecture is to introduce the general subject of population interactions by investigating a system in which two different species compete for resources. Competition between species has long been of ecological and evolutionary concern. In the early 1920s G. F. Gause, a leading Russian ecologist, formulated the *principle of competitive exclusion*. This principle states that no two populations that share the exact habitat and resource requirements may stably coexist. One will always drive the other to extinction.

The resource and habitat requirement of a species is referred to as the species' *niche*. This includes what the species eats, where it thrives, its periods of activity, where it reproduces, the temperature and humidity ranges it requires, and so on. In these terms a niche is a species "place" or "role" in the community. G. E. Hutchinson formalized this conceptual framework by describing a niche as a space or rather hyperspace defined by numerous orthogonal axes. These axes correspond to each environmental requirement or activity of a species. If we say that species A can only live in a temperature range of 12° to 30°C and a relative humidity range of 30 to 100%, and eats only grasshoppers of size 1.5 to 3.0 centimeters in length, we can draw a volume in which this species lives (Figure 15.1). If we add more axes for additional requirements, we cannot draw the volume, but we can think of it as some blob in multidimensional space.

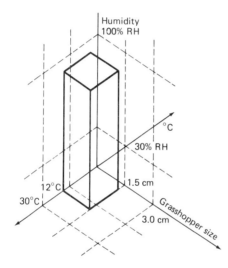

Figure 15.1 Niche volume as determined by temperature, humidity, and food size axes.

On a philosophical level, it is reasonable to assume that no two species have exactly the same niche. Let us say that we have a second species B which has the same climatic requirements as A and also eats grasshoppers. B, however, only eats a particular grasshopper whose size is 3.3 to 4.5 centimeters. Figure 15.2 shows the two niche spaces. Even though they have the same temperature and humidity requirements, the grasshopper axis separates their niches so that they can coexist.

Of course, more complex situations can and do exist, so you should be cautious not to oversimplify Hutchinson's niche concept. For example, species C has the same climate and grasshopper requirements as A, but C is active only at night while species A is active only during the day. The activity axis apparently separates their niches completely. Or does it? If species A eats up all the grasshoppers during the day, there will be no grasshoppers for species C at night. Therefore, even if the two niches are not identical, there is significant *niche overlap*.

The question that we must ask then is how much can niches overlap and

Lecture 15   Lotka-Volterra Competition Models: Competitive Exclusion   147

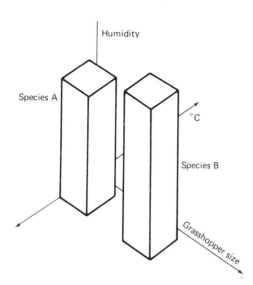

Figure 15.2 Niche volume of two species as determined by temperature, humidity, and food size axes. The two niche volumes are separated completely by the food size axis.

still allow coexistence? The model that we will use was first developed by Alfred Lotka and Vito Volterra independently during the mid-1920s, and the set of equations dealing with interspecies interaction are usually named after them. We may attempt to solve this problem of species coexistence by expanding on our basic logistic equation. The logistic equation states that individuals in a population reduce the potential growth of the rest of the population by some factor γ. In modelling competition, we assume that a second population is also reducing the potential growth of that population by some other factor. If there is no overlap in niches, that factor will be zero. If the niches are exactly the same and each individual of this second population has the same requirements as individuals of our first population, then this factor will be equal to γ. If they require more, then the factor will be greater than γ.

The important point to understand is that we are using the same rationale in both cases. The γ in the logistic equation is simply an intrapopulational competition factor. It describes the competitive effect of that population on itself. To expand this to two populations, we need only add an additional term, which describes the effect of population 2 on population 1.

The growth function for population 1 is

$$\frac{dN_1}{dt} = N_1(r_1 - \gamma_{11}N_1 - \gamma_{12}N_2)$$

where N_1 and N_2 are the population sizes of population 1 and 2, r_1 is the intrinsic rate of increase of population 1, γ_{11} is the competitive effect on population 1 by itself, and γ_{12} is the competitive effect on population 1 by population 2. Similarly, the growth function for population 2 is

$$\frac{dN_2}{dt} = N_2(r_2 - \gamma_{22}N_2 - \gamma_{21}N_1)$$

where r_2 is the intrinsic rate of growth of population 2, γ_{22} is the competitive effect on population 2 by itself, and γ_{21} is the competitive effect on population 2 by population 1.

We have two differential equations in N_1 and N_2 with coefficients r_1, r_2, γ_{11}, γ_{22}, γ_{12}, and γ_{21}. Let us repeat what it is that we want to learn from them. Our original question was, how much niche overlap can we allow and still maintain coexistence of the two species at equilibrium? Niche overlap, in terms of the two equations, can be indicated by the γ coefficients, as we discussed. In terms of population growth, equilibrium can be defined as the situation when

$$\frac{dN_1}{dt} = \frac{dN_2}{dt} = 0$$

that is, when neither population is increasing nor decreasing. We may attempt to answer our question by solving our equations for zero population growth and by comparing the γ coefficients at those points.

It would be far too difficult to solve the differential equations analytically, but we can analyze them graphically and come to some intuitively satisfying rules of thumb. We can achieve this goal by setting up a plane of all possible combinations of population sizes (Figure 15.3). It should be clear that this plane is neither a physical space or environment nor a niche space. All these elements remain constant in our model; only population sizes vary. On this plane we will plot the solutions for our equilibrium conditions. Thus for population 1, we have

$$\frac{dN_1}{dt} = 0 = N_1(r_1 - \gamma_{11}N_1 - \gamma_{12}N_2)$$

Obviously, if N_1 is zero, the population does not exist and its growth is zero; this is one solution, though not a particularly exciting one. A second set of solutions may be found by setting the quantity in parentheses equal to zero and solving. If I rearrange the equation

$$0 = r_1 - \gamma_{11}N_1 - \gamma_{12}N_2$$

as

$$N_1 = \frac{r_1 - \gamma_{12}N_2}{\gamma_{11}}$$

you will see that this is a linear equation in N_1 and N_2 and its solution set is defined by a line. To draw this solution line, I only need two points. Setting N_1 equal to zero, I get a solution at

$$N_1 = 0 \qquad N_2 = \frac{r_1}{\gamma_{12}}$$

Setting N_2 equal to zero, I get another solution at

$$N_1 = \frac{r_1}{\gamma_{11}} \qquad N_2 = 0$$

(r_1/γ_{11} is the carrying capacity for population 1.) In other words, when N_2 is equal to zero, our model reduces back to the logistic equation as it should.

Lecture 15   Lotka-Volterra Competition Models: Competitive Exclusion   149

I have plotted these two points and drawn the straight line between them (Figure 15.3). This line represents all possible nonzero combinations of N_1 and N_2 that will result in zero growth for population 1. Since population 1 on this line will not change in size, the line is called an *isocline* (iso = equal). Think of this isocline as being a carrying capacity which is contingent on the population size of the competitor. Any combination of population sizes above this line will result in a negative growth of N_1; any combination below this line will result in a positive population growth. This is indicated by the arrows next to the N_1 axis in Figure 15.3.

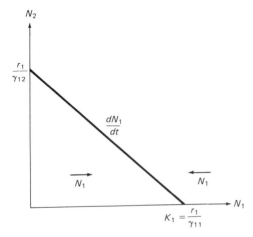

Figure 15.3 Isocline of species 1 as plotted in the plane of two species' population sizes.

The equilibrium equation for the growth of population 2 can be solved in a similar manner. Again, if N_2 is equal to zero, all values of N_1 will be possible solutions. By rearranging the expression in parentheses, we have

$$N_1 = \frac{r_2 - \gamma_{22} N_2}{\gamma_{21}}$$

The solution line is defined by the two endpoints

$$\left(\frac{r_2}{\gamma_{21}}, 0\right) \quad \left(0, \frac{r_2}{\gamma_{22}}\right)$$

I have plotted this line and indicated the direction of population growth of N_2 in Figure 15.4.

To determine the equilibrium solutions to the simultaneous growth equations, I need simply to plot both isoclines on the same coordinate plane. Obviously, the relationship between the lines will determine the solutions. There are four such possible relationships, as plotted in Figures 15.5, 15.6, 16.1, and 16.2.

In Figure 15.5 the isocline for population 1 is always greater than that for population 2. These two lines divide the space into three sections. In the area marked A above both lines both populations are above their contingent carrying capacities. Both populations have negative growth, that is, both

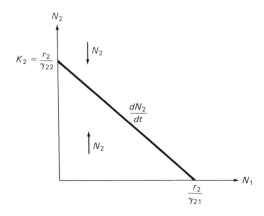

Figure 15.4 Isocline of species 2 as plotted in the plane of two species' population sizes.

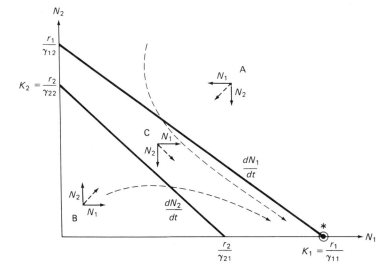

Figure 15.5 Competitive interaction where the isocline of species 1 is always greater than that of species 2.

decrease in size. This is indicated by the vectors pointing to the central area. In the area below both equilibrium lines, B, both populations are growing. This, too, results in a composite vector leading to the central area. The central area C of the plane is above the equilibrium line of population 2, but below that of population 1. Therefore, all combinations of population sizes here will show an increase in N_1 and a decrease in N_2, as indicated by the vectors. In this region N_2 will continue to decrease until it goes extinct ($N_2 = 0$). N_1, in the absence of N_2, will grow to reach its carrying capacity

$$\frac{r_1}{\gamma_{11}} = K_1$$

Lecture 15   Lotka-Volterra Competition Models: Competitive Exclusion   151

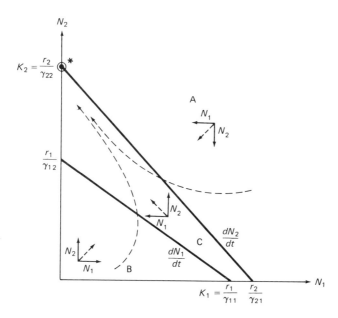

Figure 15.6 Competitive interaction where the isocline of species 2 is always greater than that of species 1.

There is then a general tendency to drive the community composition to

$$\left(\frac{r_1}{\gamma_{11}}, 0\right)$$

Thus no matter what combinations of population sizes we begin with, population 1 will "outcompete" population 2 and drive it to extinction. The stable equilibrium point of the system is

$$\left(\frac{r_1}{\gamma_{11}}, 0\right)$$

There are, of course, two unstable equilibrium points. The first is at the origin (0, 0) and the second at the carrying capacity of population 2 alone:

$$\left(0, \frac{r_2}{\gamma_{22}}\right)$$

However, as in our genetics models, the system will move away from the unstable points if there is any movement onto the plane away from them. If we are at the origin, the introduction of one individual from either species (ignoring the Allee effect!) will drive the system away from this origin. If we start at either of these two unstable points and introduce one individual of population 1, the system will move to the stable equilibrium point

$$\left(\frac{r_1}{\gamma_{11}}, 0\right)$$

The second situation, Figure 15.6, is completely analogous to our preceding example. The three equilibrium points are again

$$(0, 0) \quad \left(\frac{r_1}{\gamma_{11}}, 0\right) \quad \left(0, \frac{r_2}{\gamma_{22}}\right)$$

This time the contingent carrying capacity of population 2 is always above that of population 1: There is always an area, C, in the community plane that the system will tend to and in which population 2 will increase while population 1 decreases. This time N_1 will continue to decrease to zero, and N_2 will reach its carrying capacity

$$\left(\frac{r_2}{\gamma_{22}}\right) = K_2$$

The stable equilibrium point is

$$\left(0, \frac{r_2}{\gamma_{22}}\right)$$

The following are unstable equilibrium points:

$$(0, 0) \quad \left(\frac{r_1}{\gamma_{11}}, 0\right)$$

The arguments presented were historically the basis for the principle of competitive exclusion. Two species utilizing the same resources cannot coexist. This principle has generated a wealth of literature. Included in this literature are discussions of possible tautologies in the definitions of niche and species. Two species coexisting must not share the same niche. Two species sharing the same niche cannot coexist. This leads to the experimental problem of determining the niche requirements independently of the fact of coexistence.

Species having apparently similar niches are found to coexist in nature. What maintains this coexistence? How much similarity is allowed? Lecture 16 will address these issues.

lecture 16

Lotka-Volterra competition models: coexistence

We have developed a model to describe competition between species. The basis for this model is the logistic growth equation with the addition of interpopulation density dependence. The first two cases that we have analyzed gave the conditions of the competitive exclusion. However, similar species do coexist in nature. Let us consider, therefore, conditions of coexistence within the same class of Lotka-Volterra models that we developed in the previous lecture.

We have considered two cases in which one species always outcompeted the other. Graphically, this meant that the isoclines never intersected in the positive quadrant of the community plane. Figures 16.1 and 16.2 present intersecting isoclines. Because the intersection point is on both zero growth lines, it too must be an equilibrium point, that is, at this point both populations will maintain a constant population size. Thus there is a possibility of reaching coexistence of the two populations at equilibrium. The question is whether this is a stable or an unstable equilibrium. We may answer by analyzing the population growth vectors in each area of the plane. The two isoclines now divide the plane into four such areas.

Study Figure 16.1. The areas labeled A and B have properties that are identical to the similarly labeled areas in Figures 15.5 and 15.6. In area A both populations are above their contingent carrying capacities and so must *decrease* in size. In area B they are both below their contingent carrying capacities and must *increase* in size. If the population sizes are driven to the intersection point

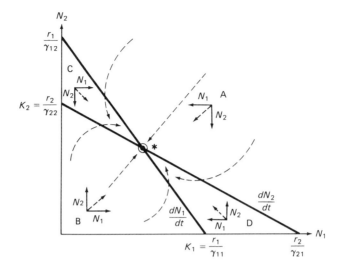

Figure 16.1 Competitive interaction where the isoclines of the two species intersect and where a stable coexistence is possible.

of the lines, they will remain there, since this is an equilibrium point. What happens if the population sizes move away from this point by chance, or if the combined populations move directly into areas C or D from A or B? In area C population 1 is below its isocline, so it will increase. Population 2 will decrease, however, because it is above its isocline. The combined population growth vectors point toward the central equilibrium. In area D population 1 must decrease and population 2 must increase because of the positions of their isoclines. Again, the combined population growth is toward the center. Thus no matter where we start in the population plane, the general tendency is toward the central equilibrium point. It then is the stable equilibrium point for the system, whereas

$$(0, 0) \quad \left(\frac{r_1}{\gamma_{11}}, 0\right) \quad \left(0, \frac{r_2}{\gamma_{22}}\right)$$

are unstable equilibrium points. Thus the two populations may maintain nonzero population sizes at equilibrium. This case describes a competition system in which populations may *stably coexist*.

Situation 4 is somewhat different (Figure 16.2). Areas A and B have the same properties as in all the previous examples. Again, the intersection of the isoclines is an equilibrium point. Is it a stable equilibrium? To answer this, let us again look at areas C and D. Area C is characterized by being above the population 1 isocline and below the population 2 isocline. Thus population 1 will always decrease and population 2 will always increase in this area. The composite growth vector is directed toward the equilibrium point

$$\left(0, \frac{r_2}{\gamma_{22}}\right)$$

Lecture 16   Lotka-Volterra Competition Models: Coexistence   155

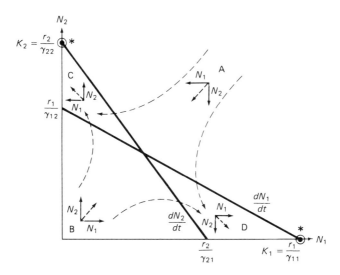

Figure 16.2 Competitive interaction where the isoclines of the two species intersect and where an unstable coexistence is indicated.

Thus in area C, population 2 outcompetes population 1 and drives it to extinction. Area D is a mirror reflection of area C. Population 2 is now limited in growth while population 1 increases. The vector directs the combined populations to the point

$$\left(\frac{r_1}{\gamma_{11}}, 0\right)$$

Population 2 is excluded by population 1. Any movement from the central equilibrium point into area C or D will result in the exclusion of one or the other populations. Thus only *unstable coexistence* is possible in this situation; either population 1 or 2 will outcompete the other population. The final result depends on which area, C or D, the community moves into, and this in turn depends on the initial combination of population sizes.

Given the parameters $r_1, r_2, \gamma_{11}, \gamma_{22}, \gamma_{12}$, and γ_{21}, you should be able to reconstruct one of these four plots and tell whether there is stable coexistence, unstable coexistence, clear competitive exclusion of species 1, or clear competitive exclusion of species 2. By determining the endpoints

$$\left(0, \frac{r_2}{\gamma_{22}}\right) \quad \left(0, \frac{r_1}{\gamma_{12}}\right) \quad \left(\frac{r_1}{\gamma_{11}}, 0\right) \quad \left(\frac{r_2}{\gamma_{21}}, 0\right)$$

you can determine if there is an internal equilibrium point; if not, you can determine which population will be excluded. If there is an internal equilibrium point, that is, there is coexistence, is this point stable?

There should be a way to make this decision based on logic alone without the aid of drawing little arrows on the graph. Not surprisingly, there is such a method.

Look back to Figure 16.1, the diagram of a system with stable coexistence. From the intersections on the N_1 and N_2 axes, it is clear that

$$\frac{r_1}{\gamma_{12}} > \frac{r_2}{\gamma_{22}} \quad \text{and} \quad \frac{r_2}{\gamma_{21}} > \frac{r_1}{\gamma_{11}}$$

Because all the coefficients are positive, I can make the following simple rearrangements.

$$r_1\gamma_{22} > r_2\gamma_{12} \qquad r_2\gamma_{11} > r_1\gamma_{21}$$
$$r_1\gamma_{22}\gamma_{11} > r_2\gamma_{12}\gamma_{11} \qquad r_2\gamma_{11}\gamma_{12} > r_1\gamma_{12}\gamma_{21}$$

It is clear then that

$$r_1\gamma_{22}\gamma_{11} > r_1\gamma_{12}\gamma_{21}$$

or

$$\gamma_{22}\gamma_{11} > \gamma_{12}\gamma_{21}$$

From this we may conclude that, for a stable coexistence to occur, it is necessary that the product of the intrapopulation competition coefficients be greater than the product of the interpopulation competition coefficients. Alternatively, we see from Figure 16.2 that the condition of competitive exclusion is

$$\gamma_{12}\gamma_{21} > \gamma_{11}\gamma_{22}$$

To summarize, if populations limit their own growth more than they do each other's, the populations will coexist. If populations limit each other more than they limit themselves, there will be competitive exclusion.

This all makes intuitive sense. However, a problem still remains. The coefficients are summary values, and only in terms of these summary values have we made an attempt to predict whether or not coexistence is possible. Let us attempt to interpret these results in terms of resource utilization.

Assume that we have two species, each competing for a common limiting resource. Suppose that the level of resources available limits the growth of both populations. I will write the equations in the following form:

$$\frac{dN_1}{dt} = N_1[r_1 - \alpha_1 f(\text{resource})]$$

$$\frac{dN_2}{dt} = N_2[r_2 - \alpha_2 f(\text{resource})]$$

The variables N_1 and N_2 and parameters r_1 and r_2 are the same as we used previously. The α terms are coefficients of efficiency of resource usage for each species and f(resource) is a function that describes the total available resource. A simple yet reasonable approach to this resource function would be to assume that each species is taking up a certain amount of the total resource T at a rate of β_1 and β_2 per individual, respectively. Our resource function can then be written

$$f(\text{resource}) = T - \beta_1 N_1 - \beta_2 N_2$$

Lecture 16 Lotka-Volterra Competition Models: Coexistence

I can insert this function into our equation and solve for the isoclines

$$\frac{dN_1}{dt} = 0 \quad \frac{dN_2}{dt} = 0$$

You should be able to prove that the isoclines are defined by

$$\text{Species 1}: N_1 = \frac{-\beta_2}{\beta_1} N_2 + \frac{\alpha_1 T - r_1}{\alpha_1 \beta_1}$$

$$\text{Species 2}: N_1 = \frac{-\beta_2}{\beta_1} N_2 + \frac{\alpha_2 T - r_2}{\alpha_2 \beta_1}$$

You should immediately notice that if the slopes of the lines are identical $(-\beta_2/\beta_1)$, the isoclines are parallel. This situation falls into the category of competition described in Figure 15.5 or 15.6. Stable coexistence is impossible. One population must outcompete the other. We can conclude that if one resource is the limiting element in the competitive system described by the preceding model, we will see competitive exclusion.

Under what conditions might our model be incorrect? First, there may be a second resource that limits one or both of the populations. Such secondary limitations may force an intersection of the isoclines to give an equilibrium on the plane. For example, two bird species may feed on, and be limited by, the same set of insects, but one species may be additionally limited by the number of suitable nesting sites. Second, there may be developmental constraints that produce nonlinearity of resource utilization. The quantity or quality of resources used may change between juvenile and adult stages, for instance. Third, the density dependence itself may be highly nonlinear. Isoclines that are strange, wavy lines may intersect at more than one point (Figure 16.3). All these cases are possible, though it is generally accepted that more than one limiting resource must be occurring to assure coexistence.

I have tried to develop a model starting with some simple assumptions and leading to some not so simple conclusions. We began by assuming that two interacting populations or species, when living in the absence of the other, grow according to a density-dependent, logisticlike function. When living together in the same locality, however, the two species compete to some extent for similar resources. This competition then reduces the rate of growth in both populations.

The important assumption that we are making, though, is that this reduction is similar to the density-dependent growth regulation that each population asserts on itself in logistic growth. This being the case, we simply add another term to our logistic equation. Our competition equation can then be written schematically:

Growth = Logistic equation − Competitive reduction

What are our conclusions from this rationale? The first is that if there is no intersection of the growth isoclines (that is, there is no nonzero root to

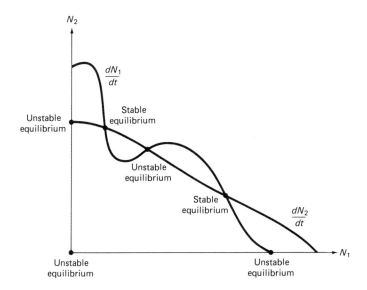

Figure 16.3 Competitive interaction where the isoclines of both species are nonlinear and where several stable and unstable coexistence points are indicated.

the two simultaneous equilibrium equations), then the species with the isocline farthest from the origin will always dominate the competitive interaction. It will always exclude or drive the other population to extinction. This is particularly true when only one resource is limiting. When two or more resources are limiting, the isoclines may intersect. This intersection implies the possibility for coexistence of the two species, even though they are competing. The factor determining whether this coexistence is stable and persistent or unstable and thus ephemeral is the relative strength of our density-dependent coefficients. If the internal density controls are greater than the competitive controls, then the species may coexist. If the populations exert more pressure on each other than on themselves, the coexistence is unstable and either species may win, with the final victor being determined by initial population sizes.

No good tale is complete without its homiletics. Species do compete, and yet they often coexist. You would then suppose from our model that more than one resource is limiting. This may be the case. You should, however, be aware that there may be other factors involved. For example, the environment may not be uniform, as is often the case. In such an instance the two species may compete differently or have different r coefficients in different parts of the environment. They may then coexist due to a trade-off in competitive abilities. Similarly, the environment may not be uniform in time. This also requires a more complex model to describe the species interactions, and one limiting resource may still allow coexistence. Finally, there may be interactions with other species in the community such as predators. A predator, by harvesting the superior competitor, may reduce the population size of that competitor and allow for its stable coexistence with other, less effective species.

The analysis of such situations requires more complex models than I am able to present within the limited scope of this course. We will touch on the subject of more complex ecosystem models again in Lecture 18.

lecture 17

Predation

Predation or parasitoid systems differ distinctively from competition systems. In competition the response of a population to a change in population size of the competitor will be monotonic. Any increase in population size of one species will decrease the growth rate of the competitor. This decrease, in turn, may allow for a further increase in the first species which further reduces the second, and so on. These unidirectional responses continue until either one population is competitively excluded or intrapopulation density effects stop population growth, allowing for coexistence. Predator-prey interactions, on the other hand, are not unidirectional. A predator species will increase in population size when there is an abundance of food (prey) available. This increase then puts more pressure on the prey population and reduces it in size: The more mink around catching ducklings, the fewer ducks there will be the next season. This reduction in the prey population means less food for the predator, which results in an increase in mortality or a decrease in fecundity or both. The predator population decreases, thus reducing the pressure on the prey species. It then increases and the cycle starts anew.

The oscillatory pattern of population sizes has long been recognized from natural history studies, although it is not clear whether or not the prey-predator mechanism is always responsible. The most well-known example is that of the snowshoe hare and the lynx. Population sizes were estimated from trapping records kept by the Hudson Bay Company since the middle of the nineteenth century. The results indicate large oscillations with the peaks of the two populations slightly out of synchrony and with a period of about 10 years. Another set of data is that of fish populations in the Mediterranean. The story goes that not much research could be undertaken in Italy during World War I. Among the few things one could do was either go to the fish market or drive an ambulance. The famous Italian mathematician Vito Volterra and his collaborator (and son-in-law), Umberto D'Ancona, decided to do the former, leaving the ambulance driving for Hemingway. They recorded catch sizes by species over a number of years and they were able to detect an oscillating pattern with a periodicity of about 4 years. These data and the tenacity required to obtain them stimulated the work that led to the basic predator-prey model. These models, like the competition models, were developed independently by Lotka in the United States and Volterra in Italy and are generally known as Lotka-Volterra predator-prey models.

The models recall our knowledge of Malthusian and logistic growth equations and are, in essence, simple extensions of them. The first assumption that we will make is that the prey population is growing exponentially without any self-imposed density controls. However, the population is being nibbled on by each predator at a constant rate. The reasoning behind this assumption is that we want the prey population to be controlled mainly by predation pressures and not by internal density dependence. The per capita growth of the prey then is the intrinsic rate of growth minus a per individual potential of being eaten. This potential is defined by the number of predators stalking about and by the predators' efficiency, which for sake of consistency we will call γ_{12}. Thus the growth equation for the prey population, population 1, is

$$\frac{dN_1}{dt} = N_1(r_1 - \gamma_{12}N_2)$$

where r_1 is the intrinsic rate of growth of the prey population, N_1 and N_2 are the two population sizes, and γ_{12} is a predation rate.

The predator population growth equation is different from any of our previous growth models. In the exponential or logistic equation r is given as a summary coefficient describing the maximum growth under unlimited resources. Since the resource here is the prey population and it is not unlimited, we want our predator growth to be directly dependent on prey number. Thus predator growth should be directly proportional to prey population size times some constant, γ_{21}, which is the value of each prey item. The γ_{21} then describes a positive interaction that can be thought of as a conversion factor: how many chickens make a fox. The only other consideration we must make is that without the prey (chickens), the predator (fox) must die. We will follow convention and use $(-r_2)$ as this intrinsic death rate. The predator growth equation is thus

$$\frac{dN_2}{dt} = N_2(\gamma_{21}N_1 - r_2)$$

The next step is to try to solve this system of equations to see if there is a nonzero equilibrium point. As in the last lecture, we are again asking whether or not this two-species interaction can lead to coexistence. This time, however, the solutions are much easier to find. I will set the terms in the parentheses of both equations equal to zero (the equilibrium state) and solve for N_1 and N_2.

$$\frac{1}{N_1}\frac{dN_1}{dt} = 0 = r_1 - \gamma_{12}N_2$$

$$\frac{1}{N_2}\frac{dN_2}{dt} = 0 = \gamma_{21}N_1 - r_2$$

$$N_2^* = \frac{r_1}{\gamma_{12}}$$

$$N_1^* = \frac{r_2}{\gamma_{21}}$$

There are two striking features of these equilibrium values that are immediately apparent. First, there is always an equilibrium value in the positive quadrant of the species number plane. To put it more simply, according to these equations, a predator-prey interactive system always has a possible point of coexistence. This is in contrast to the competitive systems where a point of coexistence, be it stable or unstable, may not necessarily exist.

Second, the equilibrium population size of each population is directly proportional to the intrinsic rate of growth of the other population. Yet it is totally independent of its own intrinsic rate of growth. This is a curious conclusion. Imagine a harsh, resource-depauperate habitat in which our predator and prey populations are living. As we discussed in Lecture 14, r reflects an interaction of the population and environment. If our environment is depauperate, r_1, the intrinsic rate of increase of the prey species, can be quite depressed. Yet according to our model, we will see this depression in terms of low predator numbers, rather than in a small prey population size. The reason for this is that we are assuming that the prey population is controlled only by the predator; there are no intrapopulational density-dependent controls. The predator population, on the other hand, is completely dependent on the amount of prey it can harvest. The quantity harvested, if it is to be at all stable, will depend on how fast the population can reproduce. Therefore, the faster the prey reproduces, the more there is for the predators to eat. Contrarily, the prey population may be large, yet a slow reproductive rate would only support a small predator population.

As unrealistic as this conclusion may appear, there are some natural systems that seem to have a similar response. Oceanic systems have a very low standing crop of unicellular algae on which the rest of the ecosystem is based. It is a generally accepted ecological rule of thumb that the efficiency of transferring energy from one organism to another by predation is only 10%. Therefore, to support a chain of consecutive predators and prey, the volume of plant material, which is the basic link in this chain, must be large. The base is large in most terrestrial systems. The oceans, however, do not have a large base and, indeed, there is a smaller standing crop of plants than of the animals that directly eat them. Yet since the planktonic algae generally exist in low concentrations, their population growth is essentially not under density-dependent

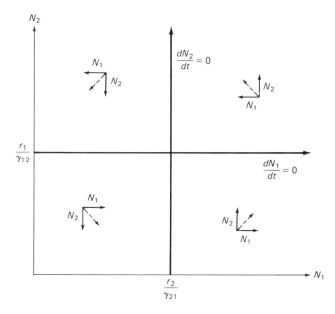

Figure 17.1 Isoclines of prey (species 1) and predator (species 2) populations plotted on the plane of two species' population sizes.

effects though such effects do exist. Their intrinsic rate of increase, r, is very large. Thus they can support a large quantity of predators.

Another curious aspect of this model is that the two equilibrium lines are dependent on one population alone. The prey isocline is defined by a predator population size and is independent of its own size; the predator isocline depends in a similar way on the size of the prey population. The isoclines are then drawn to be perpendicular to the N_1 and N_2 axes (Figure 17.1), dividing the plane into four quadrants. Furthermore, because of the signs of the interaction terms, the response of population growth to population size is not as straightforward as in the competition equations. An increase of predator population size in the area above the line causes a decrease in prey population size. Similarly, a predator population that is smaller than the equilibrium size will allow the prey population to increase. This is signified by the vectors in Figure 17.1. In counterdistinction to the competition equations a prey population size above the equilibrium value means more food, and thus a growth in N_2. Below the equilibrium value, the predator population will decrease in size. Thus to the right of the isocline the dN_2/dt vectors point up, and to the left, the vectors point down (Figure 17.1).

The composite vectors of two species thus do not point toward or away from the equilibrium point. They indicate instead a trajectory that orbits around the equilibrium (Figure 17.2). Volterra proved that this orbit is *neutrally stable*,

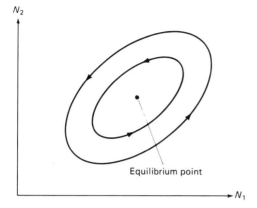

Figure 17.2 Two trajectories of predator and prey population sizes orbiting around the equilibrium point as plotted in the plane of two species' population sizes.

in that any original combination of population size will initiate a cyclical, though not circular, pattern of population sizes. This pattern will repeat itself until there is some perturbation. Such a perturbation may move the system into a new orbit. Conceptually, this is similar to the neutral stability of a Hardy-Weinberg population. There the gene frequencies remain constant until some outside force moves them to another frequency at which point they then remain at these new frequencies. Likewise, in this system the population sizes will not return to the same trajectory, but will begin orbiting in a new path (Figure 17.2).

One interesting extension of this model deals with the use of pesticides on a neutrally stable predator-prey system. Let us say that two farmers are neighbors, grow the same crop, and have the same crop pests. Farmer A applies an insecticide to his field at the peak of the pest outbreak (see Figure 17.3). Not knowing that his pest is normally controlled by a predator, he has knocked them into a smaller oscillatory cycle. Over time, he sees the population of pests continues to decline. He brags about his success to his neighbor, B, who subsequently sprays his field. He also decreases his pests, but unknowingly introduces them into a more distant orbit (Figure 17.3). He has effectively reduced the minimum number of predators that his field will support. The cycle returns to its peak and he is left with more pests than when he started. The important thing to notice is that both farmers are seeing the same average number of pests over time, r_2/γ_{21}, but they have changed the amplitude of oscillations around that number in different ways.

The satisfying result of this model is that when the population sizes are plotted against time, we see an oscillatory pattern similar to real field studies with the peaks of the two populations happily out of synchrony (Figure 17.4). Yet the neutral stability of the system is problematic and to a large extent unreal. The real world is not constant; perturbations in the guise of hurricanes, lightning, and falling rocks abound. Thus if the real systems were neutrally stable, oscillatory patterns in population sizes would never be noticed since the populations would be constantly moving from one orbit to another. How-

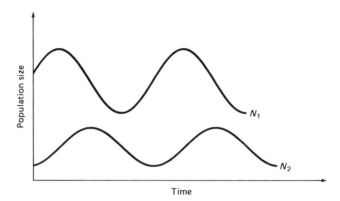

Figure 17.3 Trajectories of predator and prey populations, indicating the results of applications of insecticides at two points in time.

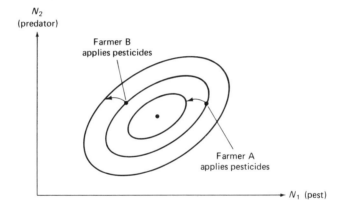

Figure 17.4 Population sizes of prey (N_1) and predator (N_2) plotted against time.

ever, after real perturbations, real populations often return to their original population pattern. What additions must we make to our model to simulate this stability pattern?

One point that is generally accepted is that most prey species have a much higher intrinsic rate of increase than their predators. It is therefore questionable whether or not prey populations can ever be completely controlled by predators alone. You can think of this in terms of how fast the predator population increases in response to a population explosion of prey. It must have an intrinsic rate of increase similar to that of the prey to catch up. If this is not the case, the prey population will probably reach a size where density-dependent intrapopulation controls come into play. To simulate this effect, we need only to introduce a logistic term to our prey equation:

$$\frac{dN_1}{dt} = N_1(r_1 - \gamma_{11}N_1 - \gamma_{12}N_2)$$

The predator equation remains the same. By determining the equilibrium point and plotting the isoclines, we get a two-species population trajectory that spirals into the equilibrium point (Figure 17.5):

$$\left(\frac{r_2}{\gamma_{21}}, \frac{r_1\gamma_{21} - r_2\gamma_{11}}{\gamma_{21}\gamma_{12}}\right)$$

Thus no matter where you start, the system always returns to the equilibrium by oscillating with continually decreasing amplitudes (Figure 17.6). The point

$$\left(\frac{r_2}{\gamma_{21}}, \frac{r_1\gamma_{21} - r_2\gamma_{11}}{\gamma_{21}\gamma_{12}}\right)$$

is a stable equilibrium point.

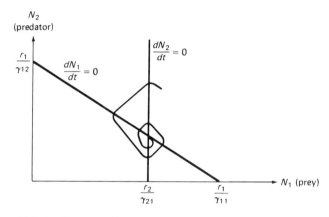

Figure 17.5 Predator-prey interaction where the prey species (1) is under intraspecific density-dependent controls. Trajectory of population sizes spirals into a stable equilibrium point.

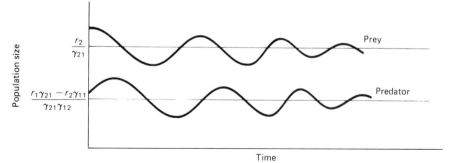

Figure 17.6 Population sizes of prey (N_1) and predator (N_2) plotted against time. The prey species population size is controlled by intraspecific density-dependent effects. The amplitude of the oscillations is dampened over time.

Another interesting addition to this model is the possibility that the prey species has a refuge where the predator cannot reach it. This is often the case when the prey is considerably smaller than the predator, or when the prey can flee to places to which the predator does not have access. An obvious example of the former case is an ecological system made up of mice, cats, and mouseholes. A system of a cat, duckling, and lake is an example of the latter. In the original Lotka-Volterra equation prey growth is limited by the number of encounters of predator and prey. This is apparent in the term $(-\gamma_{12}N_1N_2)$. If the refuge only protects a certain number, N'_1 of prey individuals, then the number of prey individuals available to the predator is $N_1 - N'_1$. If we employ this notation in the original Lotka-Volterra equations, we get

$$\frac{dN_1}{dt} = N_1 r_1 - \gamma_{12} N_2 (N_1 - N'_1)$$

and

$$\frac{dN_2}{dt} = [(N_1 - N'_1)\gamma_{21} + r_2] N_2$$

You should be able to demonstrate that the predator isocline is

$$N_1 = \frac{N'_1 \gamma_{21} - r_2}{\gamma_{21}}$$

This is a straight line, while the prey isocline is

$$N_2 = \frac{r_1 N_1}{\gamma_{12}(N_1 - N'_1)}$$

which is a hyperbola (Figure 17.7). This system also generates a dampened oscillation toward a stable equilibrium.

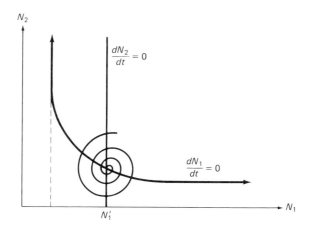

Figure 17.7 Predator-prey interaction where the prey species (1) has a refuge that protects a constant number (N'_1) against predation. Trajectory of population sizes spirals into a stable equilibrium point.

If, on the other hand, the refuge protects a constant proportion of the prey population p, the equations become

$$\frac{dN_1}{dt} = N_1(r_1) - \gamma_{12}N_2(1-p)N_1$$

and

$$\frac{dN_2}{dt} = N_2[(1-p)N_1\gamma_{21} - r_2]$$

Because we are multiplying the prey population size by a constant $(1-p)$ to signify available prey, we can introduce a new notation

$$\gamma'_{12} = \gamma_{12}(1-p)$$

and

$$\gamma'_{21} = \gamma_{21}(1-p)$$

The equations now reduce to

$$\frac{dN_1}{dt} = N_1(r_1 - \gamma'_{12}N_2)$$

and

$$\frac{dN_2}{dt} = N_2(\gamma'_{21}N_1 - r_2)$$

These are exactly like the original Lotka-Volterra equations. We can conclude that a refuge that protects a constant proportion of prey leads to neutrally stable oscillations.

Let us review the conclusions from the models up to now. A basic Lotka-Volterra model, where prey population growth is controlled by predation alone and predator population growth is facilitated by an increase in prey numbers, results in neutrally stable oscillations of both populations. If we add to this basic model density-dependent controls in prey growth, the system will oscillate to a stable equilibrium. Although not discussed, density-dependent growth of the predator will also lead to a stable equilibrium. Finally, a prey refuge, if it protects a constant number of individuals, will produce a stable predatory-prey system, but a refuge that protects a constant proportion will not.

A more general model was suggested in 1936 by the Russian mathematician Andrey Kolmogorov. He considered a system of two differential equations in the following form:

$$\frac{dN_1}{dt} = N_1 F_1(N_1, N_2)$$

$$\frac{dN_2}{dt} = N_2 F_2(N_1, N_2)$$

There are only qualitative constraints on the functions F_1 and F_2 in the model and no particular analytical expression is used. However, to reflect the character of the prey-predator interaction, the qualitative assumptions follow the properties of the simple Lotka-Volterra model. For instance, F_1 is assumed to be

N_2

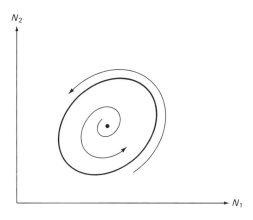

N_1

Figure 17.8 Predator-prey interaction where trajectory of population sizes generates a stable limit cycle.

decreasing with the increase of N_1 (density dependence) and N_2 (predation), whereas F_2 is assumed to be increasing with the increase of N_1 (feeding), but decreases with increasing N_2 (density dependence). Without listing all the exact assumptions made in the model, let me mention the most important result: a possibility for *limit cycle behavior*. The limit cycle is shown in Figure 17.8. It is a particular type of oscillatory pattern that is stable to any perturbations. All other trajectories tend to this one, both from inside (where an equilibrium exists but is locally unstable) and outside of the cycle. Such a special oscillatory trajectory is called a *stable limit cycle*. Its stability is an important feature which suggests that if prey-predator systems are found to oscillate in nature, the oscillations are rather of this limit cycle type than purely neutral, as in the original Lotka-Volterra model. The amplitudes of the oscillatory population sites would then remain relatively constant in spite of continuous perturbations as has been found within the few data sets available.

There is one more realistic situation that the models have been unable to reproduce. Predator populations can be so efficient that they eat all or most of the prey, causing the prey population to go extinct. Then by the consequence of starvation, the predators themselves go extinct. This situation can be taken into account with a slightly more generalized model suggested by Michael Rosenzweig and Robert MacArthur. In this model prey population growth is described by two functions. One function, $f(N_1)$, is the population growth of the prey in the absence of predators; the second, $N_2\psi(N_1)$, is the reduction of growth in the prey population caused by the predation pressure per predator individual. The predator population growth also has two elements. The first, $(-rN_2)$, is the mortality rate of the predator and the second, $kN_2\psi(N_1)$, is the benefit per prey individual caught. Thus the two equations are

$$\frac{dN_1}{dt} = f(N_1) - N_2\psi(N_1)$$

$$\frac{dN_2}{dt} = -rN_2 + kN_2\psi(N_1)$$

Notice that if $f(N_1)$ is the logistic function and $\psi(N_1)$ is linear in N_1, the new model reduces to the original Lotka-Volterra equations.

We can begin to analyze this model by noticing that the predator isocline, $dN_2/dt = 0$, is again a straight line perpendicular to the N_1 axis at a value of $\hat{N}_1$ such that

$$\psi(\hat{N}_1) = \frac{r}{k}$$

The analysis of the prey isocline is somewhat more difficult. In such a situation it is always beneficial to know in advance the results that you want. Let us assume that the prey population grows slowly in small numbers (Allee effect). This being the case, the prey population isoclines have a maximum at some nonzero population size N_1^* (Figure 17.9). If the prey population size $\hat{N}_1$ that produces zero growth of the predators is less than N_1^*, the system will follow an oscillatory pattern with increasing amplitude. Eventually this trajectory will hit the N_1 axis, implying that the predator went extinct. The prey population will then either reach its carrying capacity or the trajectory will hit the N_2 axis, implying that the prey went extinct, in which case the predators will quickly follow suit.

Another convenient feature of this model is that it can reproduce all the other results that we have found. If $\hat{N}_1 = N_1^*$, the system will be neutrally stable and will oscillate around the equilibrium. If, on the other hand, $\hat{N}_1$ is greater than N_1^*, then we have constantly decreasing oscillations and a stable

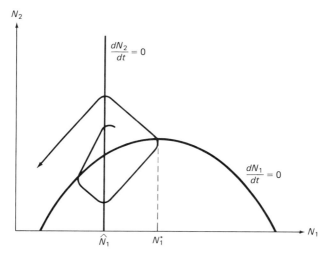

Figure 17.9 Predator-prey interaction where the predator (species 2) isocline is a straight line determined by a constant prey population size ($\hat{N}_1$) and the prey (species 1) isocline is nonlinear. The prey population size N_1^* is the value which is associated with the maximum predator size on the prey isocline. When $\hat{N}_1 < N_1^*$, the two species' trajectory spirals out from the unstable equilibrium point and intersects with the $N_1 = 0$, $N_2 = 0$, or both axes.

equilibrium. The position of $\hat{N}_1$ reflects the efficiency of the predator in catching prey. This makes intuitive sense. If the predator can effectively hunt its prey even when the prey are in small numbers, it can drive the prey extinct. This is obviously not a stable situation. If the predator is only effective at high population sizes *and* the prey population has density-dependent controls on itself, the system will be stable.

Parasitoid-host systems, in light of the predator-prey models, merit a brief discussion here. Parasitoids are parasites that have a free-living life stage (usually adult) but invade and grow in hosts in other stages (usually egg and larval). There are innumerable insects in the orders Hymenoptera and Diptera which fit such a description. The quality of interaction between host and parasitoids is the same as in predator-prey systems: The parasitoid is benefited by the host and the host is harmed by the parasitoid. The difference between these two types of systems lies mainly in the fact that the number of hosts visited by adult parasitoids will closely determine the number of parasitoids in the next generation. Models of parasitoid-host interactions are usually based on the probability of a host being parasitized. The important result is that unstable oscillatory trajectories are produced unless density-dependent controls of the host are included. Thus in spite of the biological difference the host-parasitoid and predator-prey models give qualitatively similar results.

This being the case, it may be possible to describe any ecological interaction between two species by the quality of the interactions. We can use $+$ to signify if the interaction is to the benefit of the population, $-$ if it is detrimental to the population, and 0 if it is inconsequential. In the chapter so far we have discussed $(-, -)$ or competitive interactions and $(-, +)$ or host-parasitoid, predator-prey interactions. In Lecture 18 we will look at other such interactions and consider the ways in which multispecies systems are usually developed.

lecture 18

Ecosystem models

Lecture 17 introduced the idea of characterizing species interactions by the quality of the effects of one species on another. These effects can be either positive, negative, or neutral. Using this scheme, a predator-prey interaction can be characterized by a positive effect of the prey on the predator and a negative effect of the predator on the prey. Thus we may signify the two processes of the prey, providing energy for maintenance and reproduction of the predator, and the predator, catching and killing the prey, as a simple $(+, -)$ interaction. A host-parasite interaction also falls into a $(+, -)$ category, since it has the same basic qualitative properties as the predator-prey system, in spite of the different particularities of the models.

We can categorize all possible two-species interactions using this notation system (Table 18.1). We have discussed competition and predator-prey models in some detail in Lectures 15, 16, and 17. Ignoring the case of no interaction, there are three other models. The first, *mutualism*, is perhaps the easiest to conceive. In this interaction both populations benefit from the presence of the other, so that population growth and/or the final equilibrium population sizes are increased over that of a single species in isolation. One example that readily comes to mind is the relationship between some damsel fish and sea anemones.

TABLE 18.1 Two-Species Interactions

Species 1/Species 2	+	0	−
+	$(+, +)$ Mutualism	$(+, 0)$ Commensalism	$(+, -)$ Predator-prey
0	$(0, +)$ Commensalism	$(0, 0)$ No interaction	$(0, -)$ Amensalism
−	$(-, +)$ Prey-predator	$(-, 0)$ Amensalism	$(-, -)$ Competition

Lecture 18   Ecosystem Models   173

The anemone has a system of stinging or poisonous arms that fend off attacking organisms. The damsel fish is immune to these stings and hides among the waving arms of the anemone, thereby receiving protection. Generally, the fish leave the anemone to capture food and then return to consume their catch. Pieces of food then fall onto the anemone, which consumes this delivered meal. Insect-plant pollination interactions may similarly be considered mutualistic with the plant providing the insect food or shelter and the insect performing the necessary transfer of pollen for the plant reproduction. Mutualism is of considerable ecological and evolutionary interest, possibly because of the delightful way it captures our imagination. Here, there is no conflict or trade-off; it models a beatific world. Yet on a more sophisticated level, mutualism is an interaction that must be somewhat conservative. Innovation in such interactions may easily be to the detriment of the innovator. Novel changes may disrupt the interaction, or even harm the other species and thus indirectly reduce the benefit to the innovator. Thus one-sided changes may be rejected by selection.

Amensalism and *commensalism* are more problematic. Amensalism occurs when one species harms another without deriving any benefit. I personally find this one-sided nastiness difficult to accept as a true two-species interaction. For example, if a dinoflagellate releases a secondary compound that reduces or prevents the growth of a competing species, it may also reduce the growth of a third, normally noninteracting species. The interaction between the first and third species is then amensal. Yet in essence this amensal interaction is simply a by-product of the competitive interaction. Thus in theory such an interaction could indirectly evolve from a noninteractive situation via other two-species interactions, but I cannot conceive of such a result directly evolving.

Commensalism may be a more common interaction. Here one species benefits from a second while the second species neither benefits nor is hindered by the first. However, my feeling is that this is ultimately derived from a $(+, -)$ or predator-prey interaction. If a predator or parasite evolves so as to reduce its harmful effects on its prey-host and thereby assures itself a larger and more constant resource, a commensal $(+, 0)$ interaction may evolve from a $(+, -)$ situation. Perhaps an example of this is the reduced virulence of myxoma virus in rabbits in Australia. This virus was introduced to Australia to control the rabbit population. However, over time the virus has evolved to cause fewer deaths in the rabbits, and thus provides itself with more hosts. This particular system has not reached the commensal stage, but it does demonstrate the way in which such a system may evolve. A second commensal situation, which is perhaps not evolved from a $(+, -)$ interaction, is one in which an organism provides a substrate for the growth of another. Live oaks and Spanish moss may be one example.

I will not take time here to develop models for amensalism and commensalism. It should be clear that, since one species is not controlled in any way by the second, in these types of interaction density-dependent controls are necessary. This alone may be sufficient to generate a stable commensal interaction. It may not be sufficient for an amensal model, however, in which case the second species must have density dependence in order to stabilize the system.

Mutualism is a slightly more interesting problem. One simple way to attack the problem is to model a system of compulsory mutualists. Each species needs the other to grow; in the absence of the other population each population will decline in size until it goes extinct. This is similar to the predator's dependence on the prey population in the basic Lotka-Volterra equations. We can model the mutualism interactions, then, by two predatorlike equations.

$$\frac{dN_1}{dt} = (\gamma_{12} N_2 - r_1) N_1$$

$$\frac{dN_2}{dt} = (\gamma_{21} N_1 - r_2) N_2$$

One equilibrium point of this system is clearly ($N_1 = 0$, $N_2 = 0$). A second equilibrium point exists at the intersection of the isoclines of the two populations. By solving the quantities in the parentheses, we can solve for the two isoclines.

$$\frac{dN_1}{dt} = 0 \quad \text{at} \quad N_2 = \frac{r_1}{\gamma_{12}}$$

$$\frac{dN_2}{dt} = 0 \quad \text{at} \quad N_1 = \frac{r_2}{\gamma_{21}}$$

Figure 18.1 shows the isoclines and the vector analysis of the interaction. The equilibrium point (0, 0) is clearly a stable equilibrium, whereas the point

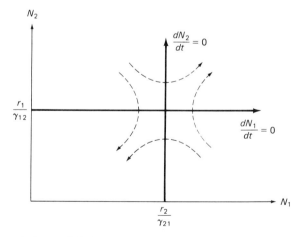

Figure 18.1 Mutualistic interaction plotted on a plane of two species' population sizes. Vectors indicate that the two species will either both go extinct or both increase infinitely.

$$\left(\frac{r_2}{\gamma_{21}}, \frac{r_1}{\gamma_{12}}\right)$$

is an unstable equilibrium. A deviation from this latter point will either lead to extinction or to constantly increasing population growth in both species.

An obvious refinement of the system would be to add a density-dependent term to the growth equation of one species and to rewrite the growth equation as

$$\frac{dN_1}{dt} = N_1(r_1 - \gamma_{11}N_1 + \gamma_{12}N_2)$$

The isocline for population 1 is now a linear equation in two variables:

$$\frac{dN_1}{dt} = 0 \quad \text{when} \quad N_1 = \frac{r_1 + \gamma_{12}N_2}{\gamma_{11}}$$

By plotting this isocline and our previous species 2 isocline in Figure 18.2, you can see that the system has not changed qualitatively. The carrying capacity r_1/γ_{11} of species 1 is now the stable equilibrium point instead of the extinction point. The equilibrium point where both species coexist is still unstable. Only when both populations are under density-dependent controls *may* a stable coexistence occur (Figure 18.3). I emphasize the word "may" since weak density-dependent controls in both species are not sufficient to assure stability. Obviously for any coexistence to occur the isoclines must intersect in the quadrant where both N_1 and N_2 are positive. Whether or not this point exists depends

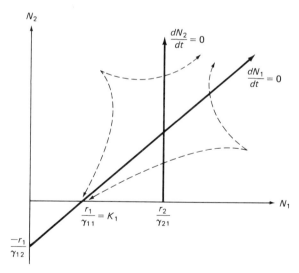

Figure 18.2 Mutualistic interaction where species 1 is under intrapopulational density-dependent controls. Trajectories point away from an unstable equilibrium coexistence point and either toward infinite population sizes for both species or toward a point $N_2 = 0$, $N_1 = K_1$.

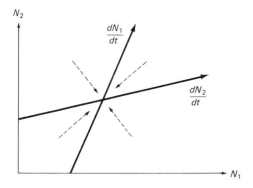

Figure 18.3 Mutualistic interaction where both species are under intraspecific density-dependent controls and a stable coexistence point exists.

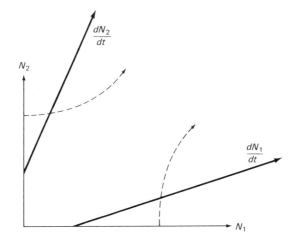

Figure 18.4 Mutualistic interaction where both species are under intraspecific density-dependent controls and there is no stable coexistence.

on the values of the six parameters $r_1, r_2, \gamma_{11}, \gamma_{22}, \gamma_{12}$, and γ_{21}. Figure 18.4 is a diagram of mutualism in which such a stable coexistence does not occur and in which both populations may increase infinitely. Interestingly enough, the stability condition for mutualism looks the same as that of competition. When the intraspecific density dependence is stronger than the interspecific interactions,

$$\gamma_{11}\gamma_{22} > \gamma_{12}\gamma_{21}$$

the coexistence point is stable.

Now that we have gone over all the two-species interactions, you should have a good introduction to the concept of ecological stability and instability. Yet I have alluded several times in these discussions to interactions involving three or more species. In Lecture 16 I mentioned that a predator feeding preferentially on an aggressive and dominating competitor may allow for coexistence of two competing species. In the example of amensalism this interaction may be a by-product of an interaction with a third species. In general, interactions among species in nature are not limited to simple two-species systems. Is it then possible to extrapolate from our simple two-species models to a multispecies community model?

The qualified answer is that, in theory, we can extrapolate our simple models to the community level. However, the qualification is that we may not be able to analyze these community models once we have generated them. To demonstrate, let us attempt to write out such a system.

Assume that we have a collection of m species, all of whose population growths may be described by general equations as follows:

$$\left. \begin{array}{l} \dfrac{dN_1}{dt} = F_1(N_1, N_2, \ldots, N_m) \\ \dfrac{dN_2}{dt} = F_2(N_1, N_2, \ldots, N_m) \\ \ldots\ldots\ldots\ldots\ldots\ldots\ldots\ldots\ldots \\ \dfrac{dN_m}{dt} = F_m(N_1, N_2, \ldots, N_m) \end{array} \right\}$$

These equations do not assume linearity and are not explicit about the strength or sign of the interactions. They simply state that the growth of a population may be dependent in some way on the population sizes of the other species.

As in the two-species interactions, we can ask whether or not there are coexistence points and whether or not they are stable. Obviously, without more specific equations we cannot solve for specific equilibrium population sizes. To get around such a problem, we may rely on the mathematician's *deus ex machina*. Assume that there exist equilibrium population sizes $N_1^*, N_2^*, \ldots, N_m^*$. Stability is the property of returning to the same point after a slight perturbation. (Remember the Fisher's theorem balloon wafting from one dimple to another?) If the perturbation is small, the change in the perturbation for each population may be determined by a linear equation. If $\epsilon_i = N_i - N_i^*$ is the deviation of population size of species i from N_i^*, then we can write, using the Taylor expansion,

$$\dfrac{d\epsilon_i}{dt} = \dfrac{dN_i}{dt} = F_i(N_1^* + \epsilon_1, N_2^* + \epsilon_2, \ldots, N_m^* + \epsilon_m)$$

$$\approx F_i(N_1^*, N_2^*, \ldots, N_m^*) + \sum_j \left.\dfrac{\partial F_i}{\partial N_j}\right|_{N^*} \epsilon_j$$

Since N^* was chosen as an equilibrium point,

$$F_i(N_1^*, \ldots, N_m^*) = 0 \qquad i = 1, \ldots, m$$

Therefore,

$$\frac{d\epsilon_i}{dt} = a_{i1}\epsilon_1 + a_{i2}\epsilon_2 + \cdots + a_{im}\epsilon_m \qquad i = 1,\ldots,m$$

where $a_{ij} = \partial F_i/\partial N_j$ is the partial derivative of the growth function F_i of species i by the population size of species j evaluated at the equilibrium. For those who are not familiar with partial derivatives, they may be thought of as the effects of the change of one variable (in our case one population size) on a function (population growth), while all other variables (population sizes) are kept constant. If the equilibrium point is stable, the value of ϵ should get generally smaller with time. Whether or not this is true depends on the values of all the a_{ij} parameters. The a_{ij} parameters can be thought of as the intensities of the ecological interactions in the community system. Indeed, the **m** × **m** matrix made up of all a_{ij} parameters of the system is called the *ecological matrix*. You should note that these matrix elements are summary parameters of r and γ-like terms, weighted by population sizes. Thus to determine whether or not a given community is stable, you must know all the intrinsic rates of increase, all the interaction terms, and the exact equations that describe the interaction. Usually, because we do not have this information, we cannot determine whether or not the system is stable.

Having said this, and having led you along this far, it is only fair to discuss those situations that we do know something about and the generalizations that have been achieved. The first such generalization is that systems involving multiple species in a predator-prey chain are stable only if the bottommost interaction, the first prey-predator system, is stable in isolation. Such systems are of general interest, since ecologists are often concerned with the energy flow from one level of consumption, or trophic level, to another. A simple example of this type of system is one with a plant on the lowest trophic level that is consumed by a herbivore on the second trophic level, eaten by a primary carnivore on the third trophic level, and finally put away by a nasty secondary carnivore, whom no one finds toothsome. Yet such simple sequential interactions may be the exception rather than the rule in nature. If the primary or secondary carnivore also eats the plant material along with its meat, the system becomes destabilized, and some species may become extinct. If, on the other hand, we have two primary carnivores competing for the same resource and the secondary carnivore is removed, the community becomes stable again.

The most extensive basic work on this problem was done by the theoretical ecologist Robert May in the early 1970s. At that time, the existing field-oriented theory stated that the more complex an ecosystem, that is, the more species and the more interactions between them, the less sensitive it will be to disturbances. Simply put, complexity leads to stability. This notion was supported by field data and common sense. If, for example, a dominant herbivore population is greatly reduced, the ecosystem may not collapse. There are carnivores eating other herbivores besides the dominant one, and there

Lecture 18 Ecosystem Models

are other herbivores eating the same plants as were being eaten by the dominant herbivore, which makes a stable structure. In a complex system there are many ways to make up for a temporary disturbance. To investigate these concepts, May developed a general model of the type used earlier and set up a theoretical community with m species. All interactions, such as mutualism, competition, predator-prey, and so on, had equal chances of occurring. Under such a system, increasing interactions between species, that is, complexity of the community, leads to increased instability. Yet, as mentioned field studies do imply greater stability with increased interspecific interaction. The conclusion that we can make from the theoretical and field results is that communities cannot be random collections of complex interactions, but that the community must be nonrandom and, in a sense, evolve through selection to stable sets of interactions.

Given the limited though significant results of community models, can whole ecosystems or global models be made? Without attempting seriously to answer this, I will acknowledge that they are made. If you are fortunate enough to see a presentation of such a model, you will be entertained with boxes, arrows, circles, and, of course, conclusions. Yet if you delve into the boxes dealing with population growth and interactions, you will find that they remain basically black. Simple two-species competition equations have six parameters to determine. A complex system has many times more. It is a basic rule of thumb of physicists that a set of equations with more than three independently derived parameters is too complex to bother solving. We ecologists have more parameters than we have pockets. To be sure, specific systems can be determined and modeled, but then we are like the Ptolomians fitting in our parameters to fit our results. As mentioned in Lecture 1, this should not be much of a surprise. We must wait for the "apple to fall" for someone to see a simpler way of dealing with these very complex problems.

Problem solving

Aquaria are more often maintained to raise and display fish than to raise and display snails. My original intention, indeed, was to breed a large number of platyfish for genetic experiments. I set up 10 aquaria and prepared them for the introduction of my experimental fish populations. While conditioning the water before the introduction of the fish, suspended algae began to bloom in each tank. Rather than chemically treating the water to control the algae, I decided to introduce *Daphnia* into the tanks. *Daphnia* are cladocerans that feed on suspended material by waving a hairy appendage as they swim through the water. The particles collect on the hairs and the animal then ingests them. The advantages of using *Daphnia* are that they are efficient in cleaning up the water in a short time and they are delectable to fish.

To introduce them into the aquaria, I collected *Daphnia* from a nearby pond. My collection turned out to be mixed with two *Daphnia* species. This really does not concern me too much since either species will serve the purpose. I put samples of the *Daphnia* mixture into my already pea-green aquaria. After a short period of time, I wound up with three tanks having only one of the introduced *Daphnia* species; all the rest had different combinations of the two species. Curious, I decided to count the *Daphnia* in every aquarium every other day (Table V.1). After 3 weeks, I plotted all the data on a two-species community plot (Figure V.1). The data points of the same aquarium are connected by arrows in chronological order. By analyzing the plotted points, what general conclusions can you make about the interaction of the two species?

> There is a clear and unmistakable tendency of the mixed species growth trajectory to swerve up and to the left. Because the scales of the two population axes are identical, a trajectory with a 45° angle would indicate that the populations are growing at the same rate. A deviation

TABLE V.1 Population Sizes of *Daphnia* Species 1 and 2 by Aquarium

	I	II	III	IV		V		Aquaria VI		VII		VIII		IX		X	
	Species	Species	Species	Species		Species		Species		Species		Species		Species		Species	
Day	1	1	2	1	2	1	2	1	2	1	2	1	2	1	2	1	2
0	82	56	68	47	61	89	85	123	43	36	115	201	13	164	80	160	110
2	102	71	83	58	74	105	103	146	52	43	139	233	16	187	96	179	131
4	126	88	102	69	90	122	123	171	63	50	167	266	19	208	114	197	155
6	153	110	124	83	109	139	147	196	75	58	200	297	22	227	134	211	182
8	183	135	150	97	131	154	174	220	90	65	237	326	27	243	158	221	212
10	216	163	181	111	157	167	205	242	106	72	279	352	31	253	184	226	246
12	251	194	217	124	186	177	240	260	125	76	326	373	37	259	214	226	284
14	285	228	259	136	219	182	278	275	147	79	377	390	43	259	247	221	325
16	319	263	306	145	257	183	320	284	171	79	433	403	51	254	283	211	369
18	351	297	359	151	298	179	366	288	198	76	491	412	60	245	323	198	417
20	379	331	418	152	344	171	416	286	228	71	552	416	69	232	366	181	468
22	404	362	481	149	393	159	468	280	262	64	614	417	80	214	412	162	521

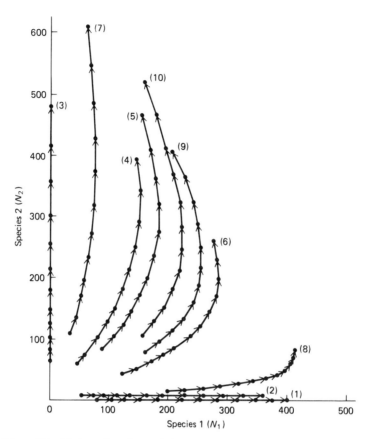

Figure V.1 Trajectories of population sizes of two species of *Daphnia* in 10 aquaria. Aquarium number is given in parentheses.

from this angle would indicate that one population is growing faster than the other. All the growth trajectories of the mixed populations deviate from a 45° angle. This, of course, is not too surprising since you would not expect two different species to have the same growth curve. Furthermore, actual population growth is dependent on population size N. Even if the two populations had the same growth parameter, we would not necessarily expect to see a 45° trajectory.

Yet there is a clear indication of interaction of the two species. Tanks 4, 5, 7, 9, and 10 have trajectories that curve back toward the N_2 axis. In the last week of growth the trajectories all have an angle greater than 90°. This indicates that species 1 is decreasing in size. You should recall that, for increasing populations below the carrying capacity, the intrapopulation density-dependent effects that we have discussed simply cause a reduction in the rate of population growth as the population increases in size. Only when a population overshoots its carrying capacity will its growth be negative, under a simple logistic growth model. There are three possible situations that would result in a decrease in one population while the other is still increasing. The first two situations assume that the two species are not sharing limited resources and are thus not competing. The first suggestion is that species 1 may be part of a predator-prey cycle and that the apparent decrease is a manifestation of the downward swing of the cycle. Alternatively, a downward trend may indicate a situation where a resource used by species 1 is not only limited but also nonrenewable. Clearly, a nonrenewable resource would result in a constantly deteriorating environment. Both K and r would be expected to decrease and the population would be reduced and eventually die. The third alternative is that the two species are competing for resources, and that population 2 is decisively reducing the growth potential of species 1 in all the aquaria.

In Figure V.2 I have plotted the growth curves of species 1 in all aquaria. In all populations population censuses are taken every second day. Even with this rather sparse data set, it seems reasonable to reject

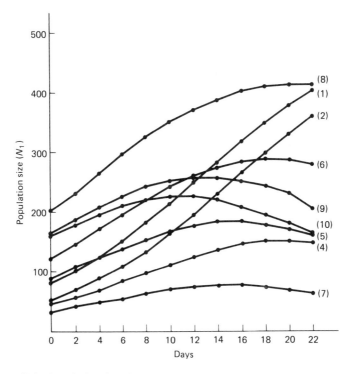

Figure V.2 Population size of *Daphnia* species 1 over time. Aquarium number is given in parentheses.

the first two hypotheses. To begin, the two monocultures, aquaria 1 and 2, have vastly different growth curves than the mixed cultures. The monoculture growth curves suggest the general sigmoidal shape of logisticlike growth. On the other hand, the mixed cultures all have a parabolic growth curve. In addition the growth rate maxima are vastly different between the populations from the mixed species aquaria and are reached at vastly different population densities, from about 50 individuals in population 7 to about 240 in population 8. It appears that the growth response is independent of population size. I would conclude that either each aquarium is completely different, in which case no comparison can be made, or the first two hypotheses are incorrect and there is some competitive interaction occurring with species 2. Because aquaria 1 and 2 give similar results, I tend to favor the second conclusion.

If species 2 is indeed reducing the growth rate of species 1, we cannot say that it is a competitive $(-, -)$ interaction unless species 1 also reduces the growth of species 2. If, for example, species 2 excretes an inhibitory substance that reduces its own growth as well as that of species 1, the interaction would be an amensal interaction with intrapopulation density-dependent controls on species 2. The density dependence would be independent of species 1 population size. If there is competition for a mutually limiting resource, such as suspended algae, there should be a depression in species 2 growth rate proportional to the population size of species 1. Plot the growth curve of species 2 and discuss the curves in this light.

Figure V.3 is a graph of the growth curves of species 2. It is immediately clear that the extent of competitive interaction by species 1 on the growth of species 2 is much less than the reciprocal interaction. Still there does seem to be some growth depression, especially in the mixed cultures 5, 6, and 9. Perhaps this is clearest in populations 5 and 9 where the growth curves intersect the monoculture growth curve. It is apparent that something besides the population size of species 2 itself is reducing its growth.

Having examined the monoculture growth curve from aquarium 3, I noticed that the rate of population growth in this tank is not slowing down. The population either does not have density-dependent growth, which is unlikely, or the density dependence is so weak that, within the range of population densities that we see, the rate of growth dN/dt is still increasing with each additional individual.

This leads to an interesting problem. We know from the Lotka-Volterra equations that the result of a competitive interaction depends on the value of the interaction terms γ. Is there a way to determine the intrinsic rates of increase

Chapter V   Problem Solving

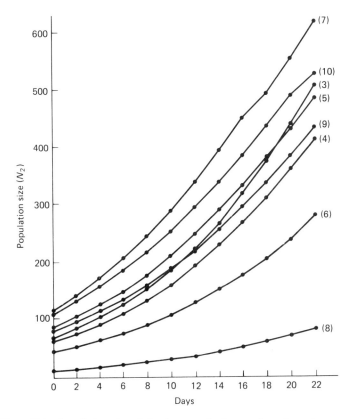

Figure V.3 Population size of *Daphnia* species 2 over time. Aquarium number is given in parentheses.

and the interaction terms of the two populations? If this is possible, can we then explain the general interactive pattern seen in Figure V.1?

I will begin my answer to this problem with a short preface. Any estimate of growth parameters or competitive interaction terms depends on the model used. Calculations based on models assuming a linear decrease in per capita growth, such as the Lotka-Volterra models, will give quantitatively different results from calculations based on nonlinear models. As such, I will explicitly state that all estimates that I will make concerning the growth parameters of the *Daphnia* will be based on the Lotka-Volterra models.

The next assumption that I must make to determine the parameters is that every aquarium provides exactly the same environment to the *Daphnia*. If this is true, I can use the data from the monocultures to estimate r_1, r_2, γ_{11}, and γ_{22} and then use these values with the data from the two aquaria to estimate γ_{12} and γ_{21}.

There are two ways that we can estimate the r and intraspecific γ parameters from the monoculture data. The first makes use of our conclusions from the problem of optimal harvesting that we discussed in Lecture 14. I know that the population size that results in the greatest population growth is $K/2$, one-half the carrying capacity. I also know that the rate of change at this population size is

$$\frac{dN}{dt} = \frac{rK}{4}$$

I can rearrange this equation and solve for r as follows:

$$r = \frac{dN}{dt}\left(\frac{4}{K}\right)$$

$K/2$ can be determined by estimating on the graph the point at which the rate of increase is at its maximum. I can further estimate dN/dt by taking two population sizes near this point and dividing their difference by the time interval. This method requires eyeballing the graph and therefore could be sloppy. Furthermore, aquarium 3 does not demonstrate the slowing down of the growth rate, and so in this case I could not use this technique at all.

The second method uses a simple algebraic manipulation of the logistic growth equation. Take population sizes at two moments of time from one population, $N^{(1)}$ and $N^{(2)}$. From the graph, it is possible to determine the change in population size until the next measurement, or $dN^{(1)}/dt$ and $dN^{(2)}/dt$. If we assume that the parameters are constant throughout the growth of the population, we can set up two simultaneous equations.

$$\frac{dN^{(1)}}{dt} = N^{(1)}(r - \gamma N^{(1)})$$

$$\frac{dN^{(2)}}{dt} = N^{(2)}(r - \gamma N^{(2)})$$

By subtracting the second equation from the first and rearranging, I can solve for γ as follows:

$$\gamma = \left(\frac{1}{N^{(1)}}\frac{dN^{(1)}}{dt} - \frac{1}{N^{(2)}}\frac{dN^{(2)}}{dt}\right)\left(\frac{1}{N^{(2)} - N^{(1)}}\right)$$

I can now solve for r by plugging this γ value back into either equation.

For example, aquarium 1 had a monoculture of *Daphnia* species 1. During the first time period, the population started at 82 individuals and increased by 20 in 2 days, or

$$\frac{dN^{(1)}}{dt} = \frac{20 \text{ individuals}}{2 \text{ days}} = \frac{10 \text{ individuals}}{\text{day}}$$

In the one before last time period recorded the population went from 351 to 379 individuals in 2 days. Thus I know that

$$\frac{dN^{(2)}}{dt} = \frac{28 \text{ individuals}}{2 \text{ days}} = \frac{14 \text{ individuals}}{\text{day}}$$

I can now solve for γ_{11} as follows:

$$\gamma_{11} = \left(\frac{dN^{(1)}}{dt}\frac{1}{N^{(1)}} - \frac{dN^{(2)}}{dt}\frac{1}{N^{(2)}}\right)\left(\frac{1}{N^{(2)} - N^{(1)}}\right)$$

$$= \frac{\left(\frac{10 \text{ individuals}}{\text{day}} \cdot \frac{1}{82 \text{ individuals}} - \frac{14 \text{ individuals}}{\text{day}} \cdot \frac{1}{351 \text{ individuals}}\right)}{(351 - 82 \text{ individuals})}$$

$$= 0.0003 (\text{day} \cdot \text{individuals})^{-1}$$

Similarly, I can solve for γ_{22} from the data of aquarium 3.

$$\gamma_{22} = \frac{\left(\frac{15 \text{ individuals}}{2 \text{ days}} \cdot \frac{1}{68 \text{ individuals}} - \frac{59 \text{ individuals}}{2 \text{ days}} \cdot \frac{1}{359 \text{ individuals}}\right)}{(359 - 68)}$$

$$= 0.0001 (\text{day} \cdot \text{individual})^{-1}$$

Plugging these values back into the equations, I get

$$r = \left(\frac{dN}{dt}\frac{1}{N} + \gamma N\right)$$

$$r_1 = \left(\frac{10}{82} \cdot \text{day}^{-1} + 0.0003 \cdot 82 \text{ day}^{-1}\right)$$

$$= 0.1466 \text{ day}^{-1}$$

and

$$r_2 = \left(\frac{7.5}{68} \text{ day}^{-1} + 0.0001 \cdot 68 \text{ day}^{-1}\right)$$

$$= 0.1171 \text{ day}^{-1}$$

I can determine the interaction parameters γ_{12} and γ_{21} directly from the Lotka-Volterra equations.

$$\frac{dN_1}{dt} \cdot \frac{1}{N_1} = r_1 - \gamma_{11}N_1 - \gamma_{12}N_2$$

$$\gamma_{12} = \frac{1}{N_2}\left(r_1 - \gamma_{11}N_1 - \frac{dN_1}{dt} \cdot \frac{1}{N_1}\right)$$

Similarly,

$$\gamma_{21} = \frac{1}{N_1}\left(r_2 - \gamma_{22}N_2 - \frac{dN_2}{dt} \cdot \frac{1}{N_2}\right)$$

For my example, I will use data from the first time period of aquarium 6. Of course, to get a better estimate, I should take the average value from all time periods in all aquaria.

$$\gamma_{12} = \frac{1}{43}\left[0.1466 - (0.0003)(123) - \left(\frac{23}{2}\right)\left(\frac{1}{123}\right)\right](\text{day}\cdot\text{individuals})^{-1}$$

$$= 0.0004(\text{day}\cdot\text{individuals})^{-1}$$

$$\gamma_{21} = \frac{1}{123}\left[0.1171 - (0.0001)(43) - \left(\frac{9}{2}\right)\left(\frac{1}{43}\right)\right](\text{day}\cdot\text{individuals})^{-1}$$

$$= 0.0001\ (\text{day}\cdot\text{individuals})^{-1}$$

To determine whether or not there is an equilibrium point allowing coexistence, we must determine whether or not the isoclines of the two species intersect. Let us look at the relative positions of the endpoints of the two isoclines. The N_1-axis endpoint of species 1 isocline is $(r_1/\gamma_{11}, 0)$, while the N_2-axis endpoint is $(0, r_1/\gamma_{12})$. Similarly, the two endpoints for the species 2 isocline are $(r_2/\gamma_{21}, 0)$ and $(0, r_2/\gamma_{22})$. These four points are (489, 0), (0, 367)(species 1 isocline) and (1171, 0), (0, 1171)(species 2 isocline). It is clear that both species 1 isocline endpoints are below those of species 2. Therefore, the species 1 isocline is always below that of species 2. We should expect that species 2 will competitively exclude species 1 in all mixed culture aquaria. Both the general interpretation and the relative position of the species 1 isocline are consistent with the data in Figure V.1.

My interest in the fish is swinging away from genetics to the ecology of the aquaria. Suppose I introduce the fish into the mixed culture aquaria. What effect will the fish predation have on the competitive interaction of the *Daphnia* species?

Without knowing either the feeding preferences of the fish or the consumption rates, it is impossible to make a quantitative prediction. However, several qualitative predictions can be proffered depending on the proposed scenarios.

For example, suppose that the platyfish have no preference for either species of *Daphnia* but eat each at a rate that is proportional to their occurrence in the community. If the fish were introduced into the aquarium before species 2 excluded species 1 by competition, the predation could push the two-species composition into the area of the multispecies space (notice that we now have three species axes), that is below both isoclines. Thus predation could make stable coexistence possible. This is particularly true when the predation on a species is proportional to its frequency of occurrence. The predation pressure is not constant, but lessens as the prey population becomes rare. This allows for the recovery of that species.

If, in contrast, the fish feed on only one species, coexistence of the *Daphnia* may not occur. If the platyfish only feed on species 1, the species' extinction will be accelerated. If the platyfish only feed on species 2, and the equilibrium population size of that *Daphnia* species still falls in the space above the species 1 isocline and below the species 2 isocline (comparable to area C in Fig. 15.6), species 1 will still be driven to extinction. Only if the species 2 population is driven below the species 1 isocline, will coexistence result.

These scenarios deal only with the *Daphnia*. The platyfish population will also be affected by the community composition. If the fish only consume one *Daphnia* species, its population dynamics should follow the general dampened oscillatory trajectory that we discussed in Lecture 16. The oscillations would be expected to dampen quickly because of the intraspecific density-dependent controls and the competitive controls acting on the prey species.

Alternatively, if the predator feeds on both *Daphnia* species, we should expect that the fish population would be less sensitive to any fluctuations in the population size of a given *Daphnia* species. In effect any sudden reduction of one species would be compensated by an increased predation on the other *Daphnia* species. It is as if the fish population has acquired some additional stability.

Both these conclusions concerning the fish population, and the preceding conclusions concerning the *Daphnia* competitors, concur with the results from Lecture 17. Predation in general has a stabilizing effect on the community, while competition has a generally destabilizing effect. Furthermore, certain complexities of a system, such as the predator eating two prey species from the same trophic level, can lead to greater stability, that is, such a system will be less sensitive to perturbations than simple chainlike interactions. In our example, if the fish eat both *Daphnia* species, there is a greater chance of balanced coexistence than if they eat only one of the two.

Analyses of other species interactions, such as mutualism, commensalism, or amensalism, can be performed by similar considerations and techniques. Of course, one general conclusion can always be made: The more species interacting, the greater the headache. Yet even in a rich community, the chains of actual species interactions may be quite small. This property allows us to dissect ecosystems and examine their parts. Indeed, if May's theoretical conclusions are correct (Lecture 18), we should expect ecosystems to be nonrandom aggregates of limited independent interactions.

Besides the multispecies complexity on a level above that of a single population, there is the complexity below the population level. Our consideration of population genetics is clearly one type of such subpopulational complexity. Ecologically, there is also a subpopulational complexity involving rates of survivorship, reproduction, and resource demands of different ontological stages or life stages within a population. We will study this level of complexity, a further loop in our Gordian knot, in Chapter VI.

Homework exercises

Using the following form, the Lotka-Volterra competition equations are not difficult to memorize:

$$\frac{dN_1}{dt} = N_1(r_1 - \gamma_{11}N_1 - \gamma_{12}N_2)$$

$$\frac{dN_2}{dt} = N(r_2 - \gamma_{22}N_2 - \gamma_{21}N_1)$$

N and r are defined as usual. The intrapopulation competition factors are γ_{11} and γ_{22}, while γ_{12} and γ_{21} are the interpopulation competition factors. Most mistakes are made by confusing these two types of factors. Remember that the interpopulation factor is always multiplied by the population size of the competitor; otherwise, each equation is identical to the logistic equation for single population growth.

Recall that there are four possible outcomes of two-species competition (refer to Figures 15.5, 15.6, 16.1, and 16.2). Either species 1 inevitably wins, species 2 inevitably wins, they coexist at a stable equilibrium, or the winner is determined by the initial conditions. The condition

$$\gamma_{11}\gamma_{22} > \gamma_{12}\gamma_{21}$$

insures that the competitors will stably coexist. This is also the stability condition for mutualism.

For species 1 to win in competition over species 2, it is necessary for the carrying capacity of species 1 in the absence of its competitor, r_1/γ_{11}, to be

Chapter V   Homework Exercises

greater than the conditional carrying capacity r_1/γ_{21}. In addition, of course, the coexistence condition must not be valid. Thus it is not necessary to draw a phase diagram to solve a question that asks you to identify the outcome of competition. However, you may find that it helps to do so.

The simplest form of the Lotka-Volterra predation equations is

$$\frac{dN_1}{dt} = N_1(r_1 - \gamma_{12}N_2)$$
$$\frac{dN_2}{dt} = N_2(\gamma_{21}N_1 - r_2)$$

N_1 is the population size of the prey and r_1 is its intrinsic rate of growth, while N_2 is the population size of the predator and r_2 is its death rate in the absence of prey. γ_{12} is the efficiency of predators at capturing prey, and γ_{21} is the efficiency of predators at turning captured prey items into more predators.

To solve for *average* equilibrium population sizes, you must know that

$$N_1^* = \frac{r_2}{\gamma_{21}}$$
$$N_2^* = \frac{r_1}{\gamma_{12}}$$

You should have no problem understanding these equations if you simply recall that the isocline of each species is determined by the parameters from the dynamic equation for the other species.

These simple dynamic equations lead to neutrally stable limit cycles. Remember that the addition of logisticlike density-dependent terms to the prey equation results in a stable coexistence point, as does the addition of a refuge which protects a constant number of prey. The equilibrium points in these two cases can be found by referring to Lecture 17 (pp. 166–167).

1. The geneticists' fruit flies (*Drosophila*) are also favorites of ecologists who wish to model competitive interactions in the laboratory. Suppose that we have two species of fruit flies, *Drosophila melanogaster* and *D. pseudoobscura*, competing for a common medium in population cages. At a particular temperature and humidity regime, the intrinsic rate of increase for *D. melanogaster* is $r_1 = 0.20$, whereas we measure $r_2 = 0.15$ for *D. pseudoobscura*. If $N_1^* = 500$ flies at equilibrium when $N_2^* = 0$, what is γ_{11}? If $\gamma_{22} = 0.001$, what is the equilibrium population size of *D. pseudoobscura* in the absence of *D. melanogaster*?

2. In an experiment in which we start with small numbers of both species of flies together in the cages we notice that both populations initially increase, but that one of the species eventually increases to its carrying capacity and the other species

goes extinct. If $\gamma_{12} = \gamma_{21} = 0.002$, will *D. melanogaster* or *D. pseudoobscura* win in competition?

3. Suppose that I introduce new genotypes of *D. pseudoobscura* and *D. melanogaster* into the cages. This has the effect of increasing the values of γ_{22} and γ_{11} by 10% without affecting the value of γ_{12}. If γ_{21} is now 0.0002, is the increase in the limiting effects of each population on itself enough to promote stable coexistence?

4. In another competition experiment using *D. pseudoobscura* and *D. melanogaster* we change the environmental conditions by raising the temperature of the cages. As a result, r_1 is increased but r_2 is unaffected. The respective Lotka-Volterra equations are now

$$\frac{dN_1}{dt} = N_1(0.24 - 0.0004\ N_1 - 0.002\ N_2)$$

$$\frac{dN_2}{dt} = N_2(0.15 - 0.001\ N_2 - 0.0003\ N_1)$$

Will the species coexist? If not, is it possible to determine the winner?

5. A predatory rotifer preys on a species of *Paramecium*. The dynamics of this simple laboratory system can be described by the following Lotka-Volterra equations:

$$\frac{dN_1}{dt} = N_1(0.90 - 0.30\ N_2)$$

$$\frac{dN_2}{dt} = N_2(0.08\ N_1 - 8.0)$$

What are N_1^* and N_2^*, the *average* population sizes of the predator and prey?

6. Because the predators are so efficient at finding prey, we have difficulty maintaining this system in the laboratory. If we increase the size of the container, however, the rotifers have more difficulty finding prey and the system becomes easier to maintain. If the predators are now one-third as efficient at locating prey, what is N_2^*? Why is it that the system is now easier to maintain?

7. By adding a little sand to the bottom of this larger container as a refuge for the prey, we are able to produce a stable coexistence in the rotifer-*Paramecium* system. If the number of prey who are able to escape predation in the sand is $N' = 50$, what are N_1^* and N_2^*?

8. A mutualistic interaction between an orchid and a bee that pollinates it can be modeled by the following equations:

$$\frac{dN_1}{dt} = N_1(0.10 - 0.001\ N_1 + 0.02\ N_2)$$

$$\frac{dN_2}{dt} = N_2(0.05 - 0.002\ N_2 + 1.0\ N_1)$$

where N_1 is the population size of the bee and N_2 is the population size of the orchid. Will the two species coexist?

chapter VI

Demography

lecture 19
Leslie matrices

Demography is the study of single-population dynamics with specific attention given to the ages of the individuals in a population. This return to the study of single-population growth may seem tame after the heady heights of community and ecosystem modeling. I think, though, that you will find these models equally, if not more, fascinating than the previous multispecies models. Demography involves the study of the effects and importance of different age groups on the growth and development of the population. As such, it identifies potential targets of evolutionary changes, evaluates effects of age-specific predation and competition on population growth, and forecasts growth complexities such as time lags and population inertia. Even your insurance company uses demographics to determine how large a premium you should pay for your health or life insurance.

Demography is often introduced either directly after learning the basic Malthusian growth model or after completing a discussion of logistic growth. The reason for this is simple. Demography is the study of age structure in single populations and it is basically a generalization of Malthusian growth. Yet the types of multispecies models already introduced are themselves the logical extension of the basic growth models. These models view individual organisms as billiard balls that bounce around, bud off new billiard balls, and possibly knock other billiard balls of the same or different color into various pockets and off the area of interest. Demography makes a clear philosophical break with these models by identifying the ontogenic differences between individuals in the population. The billiard balls now have numbers, and the numbers signify different needs and potentials.

What do I mean by different needs and potentials? Infants, larvae, and eggs have different energy intakes, defenses, and reproductive abilities than adults. For example, human newborns have a fairly substantial appetite but little ability to satisfy that appetite independently. Their immature systems are fairly susceptible to viral and bacterial infections. These and other factors significantly reduce the chances of a newborn celebrating its first birthday when compared to the chances of a 15-year-old reaching the age of 16. For the same reason, individuals between the ages of 14 and 25 have a considerably better chance of seeing their next birthday than those between 64 and 75. Therefore, death rates are not a uniform d for the population, but differ for each age class. Populations having different age compositions, then, will have different growth curves.

Let us formalize this pattern of change in the probability of survival. As in all our previous population growth models, we will concern ourselves only with females. Let p_x be the probability that an individual of age x will survive to age $x + 1$. p_{19} is thus the probability of a 19-year-old reaching the age of 20. Further, let L_x be the probability of a newborn individual reaching age x. The relationship between L_x and p_x is simply

$$L_x = \prod_{i=0}^{x-1} p_i$$

that is, the probability of reaching x years old is the cumulative probabilities of living from one year to the next, from birth to age x. Since these individual 1-year survivorship probabilities are independent, we multiply them together to get the combined probability.

Figure 19.1 shows a schematic graph of the log of L_x, the survivorship for the human population as a function of age. The short dip in the beginning of the curve indicates a relatively high mortality rate for the first year of life. This is then followed by a long, fairly level part of the curve which indicates a high probability of survival for the greater part of the individual's life. On reaching a certain age, the probability decreases at a rather high rate.

Survivorship curves similar to the human curve are common, though by no means universal. Long-lived organisms, such as elephants, bears, and whales, share the general shape of the survivorship curve with us (Figure 19.2, curve I). Other organisms, such as blackbirds, many plants, and oysters, have mortality

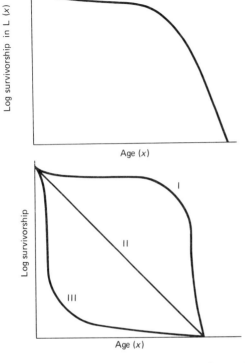

Figure 19.1 Schematic graph of the logarithm of survivorship of human populations against time.

Figure 19.2 Graph of the three basic Deevey survivorship curves.

schedules which differ significantly from this pattern, and thus do not follow such a survivorship curve. Songbirds after fledging and many herbs after germination suffer mortality at roughly constant probabilities due mainly to random, accidental causes of death. Thus, in terms of our previous notation, p_x remains constant in time. We can write this as

$$p_i = p, i = 0, 1, \ldots$$

for all i. L_x in this special case becomes

$$L_x = p^x$$

L_x is a straight line when plotted in log scale since $\ln L_x = x \ln p$. I have indicated this curve as number II in Figure 19.2. Finally, oysters, maples, and other organisms produce huge numbers of offspring or seeds and yet few of these reach maturity. The vast majority of juveniles die early in life before becoming established. However, once a seed or larva has made it, it has a good chance of living to old age. In general the survivorship curve of these organisms can be described by curve III of Figure 19.2. These three survivorship curves are generally referred to as Deevey Type I, II, and III curves in honor of the ecologist who first categorized these life history patterns.

Let us discuss the meaning of the survivorship of the first age class, L_0. The first age class is usually referred to as newborn, or newly hatched, organisms. This terminology is the result of an historical bias toward zoological studies in demography, or in population growth studies in general. Plants until very recently were not considered as individuals, as things worth counting. The last 15 years have witnessed a radical change in perspective, largely due to the work of the British plant ecologist J. L. Harper and his students. This change has led to an interesting side issue regarding the identification of the first age class. Plants, you see, have seeds that are definitely alive and that have a certain probability of remaining dormant, germinating, or dying. Thus the zero age class may be the seeds of a population or the seedlings, depending where the researcher's interests lie. In any case no matter to what it refers, L_0 by definition must be equal to one, since it is the probability of existence of something that we have already counted, and we therefore know that it must exist.

We can use our survivorship probabilities to determine how many individuals there will be in a given age class in the next year, knowing how many there are in the age class that precedes it in the current year. For example, if N_x is the number of individuals in the xth age class now, and N'_{x+1} are the number of individuals in the following age class in the next time period, then

$$N'_{x+1} = p_x N_x$$

N'_{x+1} is simply the number that survived from the previous age class.

A simple numerical example will clarify this. Table 19.1 lists the individual and cumulative survivorship values for a perennial plant by year. Assume that our study site was recently disturbed. We census plants in the site in the fall

TABLE 19.1 Survivorship Probabilities

Age Class	Year			
	0	1	2	3
p_x	0.05	0.10	1	0.0
L_x	1	0.05	0.005	0.005

and find that the numbers of individuals in the population at present are 100 seeds, 350 seedlings (1-year-old plants), 150 2-year-old plants, and 200 3-year-old plants. Next year we can expect

$100 \times p_0 = 100 \times 0.05 = 5$ 1-year-old plants

$350 \times p_1 = 350 \times 0.10 = 35$ 2-year old plants

$150 \times p_2 = 150 \times 1.0 = 150$ 3-year-old plants

$200 \times p_3 = 200 \times 0 = 0$ 4-year-old plants

Notice that we cannot predict how many seeds there will be next year. Whereas 2-year-olds must come from the previous year's 1-year-olds, newborns come from the reproductive output of all age classes. Of course, as with the survivorship curves, fecundity is not uniform throughout the population. In general newborns, seeds, or seedlings are not sexually mature and are not able to produce new offspring. Depending on the species, sexual maturity may only be reached after several years. Biennial herbs only reproduce in their second year and usually die after that year. Some trees take several years to begin reproduction, but on reaching sexual maturity will maintain a high fecundity for years to follow. In other organisms, such as humans, fecundity increases after sexual maturity is reached. A peak is achieved at some later time, and fecundity then decreases until a postreproductive stage is reached. Figure 19.3 shows different possible fecundity curves.

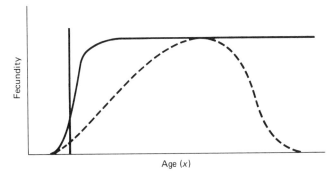

Figure 19.3 Schematic graph of three fecundity curves.

We can introduce fecundity into our previous model by defining m_x as the age-specific fecundity, that is, the number of offspring a female of age x will produce during one time period. For example, m_3 is the average number of female offspring a female will produce from age 3 to age 4. Thus if we wish to know the number of offspring that will be produced during a given time period, we simply sum all expected offspring of each individual age class. Where N_0 is the number of individuals of the zero age class, we may write this as

$$N_0 = \sum_{i=0}^{\infty} m_i N_i$$

Let us return to a numerical example. Assume that the individuals of the plant species whose survivorship we considered previously take 1 year after germination to reach reproductive maturity, produce seeds in their second and third years, and then die. Such a plant is a perennial, and its fecundity schedule is shown in Table 19.2. The number of seeds expected this year is

$$N_0 = N_1 m_1 + N_2 m_2 + N_3 m_3$$
$$= (350)(0) + (150)(200) + (200)(50)$$
$$= 0 + 30{,}000 + 10{,}000$$
$$= 40{,}000$$

TABLE 19.2 Fecundity Schedule for a Perennial

	Year			
Age Class	0	1	2	3
m_x	0	0	200	50

This number was severely reduced to 100 by the disturbance already mentioned. Barring further disturbances, we can predict the number of seeds that we should find next fall by multiplying next year's predicted age distribution by the fecundity coefficients as follows:

$$\begin{aligned}N'_0 &= N'_1 m_1 + N'_2 m_2 + N'_3 m_3 \\ &= (N_0 p_0)m_1 + (N_1 p_1)m_2 + (N_2 p_2)m_3 \\ &= (100)(0.05)(0) + (350)(0.10)(200) + (150)(1.0)(50) \\ &= 0 + 7000 + 7500 \\ &= 14{,}500\end{aligned}$$

Let me summarize what we have done so far. The size of the zero age class can be predicted by summing the potential reproductive output of each present age class, that is, by summing the products of the fecundity, survivorship, and the population size of each living age class. The sizes of the subsequent age classes are projected by the product of the survivorship parameter and the size of each previous age class. These can be written

$$\begin{aligned}N'_0 &= m_1 p_0 N_0 + m_2 p_1 N_1 + \cdots + m_n p_{n-1} N_{n-1} \\ N'_1 &= p_0 N_0 \\ N'_2 &= p_1 N_1 \\ N'_3 &= p_2 N_2 \\ N'_n &= p_{n-1} N_{n-1}\end{aligned}$$

I have purposely written these equations in such a form to foreshadow a convenient shorthand method of writing them all together. This method entails using matrix algebra, where one vector is the population census by age class, which is simply the vector of $N_0, N_1, \ldots, N_x$. For those unfamiliar with matrix algebra, multiplication of a row and column is simply

$$(a\ b\ c\ d\ e) \times \begin{pmatrix} f \\ g \\ h \\ i \\ j \end{pmatrix} = af + bg + ch + di + ej$$

Each element in the row vector is multiplied by the corresponding sequential

Lecture 19 Leslie Matrices

element going down the column vector. These products are then summed to give a single number. You should be able to convince yourself that N'_0 is simply

$$N'_0 = \begin{pmatrix} p_0m_1 & p_1m_2 & p_2m_3 & \cdots & p_{n-1}m_n \end{pmatrix} \begin{pmatrix} N_0 \\ N_1 \\ N_2 \\ \cdot \\ \cdot \\ \cdot \\ N_{n-1} \end{pmatrix}$$

Perhaps less obviously, we can write each subsequent equation for other age classes:

$$N'_i = (0\,0\,0 \ldots p_{i-1} \ldots 0) \begin{pmatrix} N_0 \\ N_1 \\ \cdot \\ \cdot \\ \cdot \\ N_{i-1} \\ \cdot \\ \cdot \\ \cdot \\ N_{n-1} \end{pmatrix}$$

We may combine all the row vectors into one matrix as follows

$$\begin{vmatrix} p_0m_1 & p_1m_2 & p_2m_3 & \cdots & p_{n-1}m_n \\ p_0 & 0 & 0 & \cdots & 0 \\ 0 & p_1 & 0 & \cdots & 0 \\ 0 & 0 & p_2 & \cdots & 0 \\ \cdot & \cdot & \cdot & \cdots & \cdot \\ \cdot & \cdot & \cdot & & \cdot \\ \cdot & \cdot & \cdot & \cdots & \cdot \\ 0 & 0 & 0 & \cdots & p_{n-1} & 0 \end{vmatrix}$$

This matrix is called the *Leslie matrix*, after the ecologist P. H. Leslie who devised this model in the mid-1940s. It consists of all the demographic parameters of a population. We will denote this matrix by **L**. The column vector containing all the numbers of individuals of each age class is the age structure vector and will be denoted by **N**. Our entire population age distribution for the next year **N**′ is then given by the equation

$$\mathbf{N}' = \mathbf{L} \times \mathbf{N}$$

Let us go back to our perennial and write the Leslie matrix for this species.

$$\mathbf{L} = \begin{vmatrix} 0 & 20 & 50 & 0 \\ 0.05 & 0 & 0 & 0 \\ 0 & 0.10 & 0 & 0 \\ 0 & 0 & 1 & 0 \end{vmatrix}$$

The age distribution for the next year is

$$\mathbf{N'} = \mathbf{L} \times \mathbf{N}$$

$$= \begin{vmatrix} 0 & 20 & 50 & 0 \\ 0.05 & 0 & 0 & 0 \\ 0 & 0.10 & 0 & 0 \\ 0 & 0 & 1.0 & 0 \end{vmatrix} \times \begin{vmatrix} 100 \\ 350 \\ 150 \\ 200 \end{vmatrix}$$

$$= \begin{vmatrix} (0)(100) + (20)(350) + (50)(150) + (0)(200) \\ (0.05)(100) + (0)(350) + (0)(150) + (0)(200) \\ (0)(100) + (0.10)(350) + (0)(150) + (0)(200) \\ (0)(100) + (0)(350) + (1.0)(150) + (0)(200) \end{vmatrix}$$

$$= \begin{vmatrix} 14{,}500 \\ 5 \\ 35 \\ 150 \end{vmatrix}$$

We now have a method of considering the differential effects of age on a population and we have a neat notation to predict the composition of the population in the next year. This is all we need for direct computations. However, we can study qualitative properties of demographic growth without doing much computation. Questions we may raise are: What is the relationship of this model to Malthusian growth? What is the advantage of this model over the simpler models used previously? Lecture 20 addresses these issues.

lecture 20
Growth rate and equilibrium age distribution

Forging ahead with our demographic growth models, let us deal directly with the problem of population growth. We have defined two sets of parameters, survivorships p_i and fertilities m_i, that together describe the growth potential for each age class in the population. By arranging these parameters in a Leslie matrix, **L**, and by listing the number of females in each class (remember that in all growth equations we are following only the females) in a vertical population vector, **N**, we can write in shorthand form the transition for numbers of individuals in each age class after one period of time as

$$\mathbf{N}' = \mathbf{L} \times \mathbf{N}_0$$

You can see that this is a repetitive process, and therefore, after two periods of time

$$\mathbf{N}'' = \mathbf{L} \times \mathbf{N}' = \mathbf{L}^2 \times \mathbf{N}_0$$

In general we have a kind of geometric series

$$\mathbf{N}_t = \mathbf{L}^t \times \mathbf{N}_0$$

This still does not answer the question of population growth per se. Yet there are not many logical jumps to be made to reach a description of growth. First, if you add up the individuals in all the age classes, you will get the total population size N. If you do the same for the next time period, you can describe growth rate as

$$G = \frac{N'}{N}$$

This is analogous to what we did when we discussed the discrete rate of growth R in Lecture 13. We have to be cautious, however, since there is a logical pitfall into which we are plummetting. The parameter R, the per capita rate of growth in the Malthusian model, describes the number of female offspring per female that we can expect after one period of time in a discrete model with *nonoverlapping generations*. Each individual produces R offspring and then dies, that is, each individual replaces itself exactly by a factor of R. As a result of this,

$$N' = RN$$

In contrast, our demographic model allows for overlapping generations. Offspring can live and/or reproduce while their parents are still alive and/or reproducing.

How then may we get a measure of population growth when there are overlapping generations? One measure would be the average number of female offspring a newborn female is expected to produce in her lifetime. This would be the number of females with which she is replacing herself. If this number is greater than one, the population is growing; if it is less than one, the population is decreasing. If it is equal to one, the population is stationary. Of course, this growth (or decrease) is basically exponential, as in our simpler models. To calculate this measure, we should sum the average number of female offspring a female produces at each age multiplied by the probability that she reaches this age. We add up all these values until she either stops reproducing or dies. The average number of female offspring produced per age class is simply the age-specific fecundity values, m_x, introduced in the last lecture. The probability of reaching a given age, L_x, was also introduced in the last lecture. Remember that this is the cumulative probability of surviving from birth to age x, not just the probability of surviving from age $x - 1$ to age x, which is p_{x-1}. We now have all the elements to our answer; we just need to put them together. Let us compute the number of female offspring a female is expected to produce over her lifetime, R_o, as follows:

$$R_o = \sum_{i=0} L_i m_i$$

Keep in mind that R_o does not equal R except in such special cases as when the population is not growing, in which case both R and R_o are one, or when only individuals of one particular age class reproduce.

To make sure that you thoroughly understand what we have done, I will refer back to the example of the perennial that we used in the last lecture. Table 20.1 reassembles the pertinent fecundity and survivorship values. We can calculate R_o for this population as follows:

Lecture 20   Growth Rate and Equilibrium Age Distribution

$$R_o = \sum_{x=0}^{3} L_x m_x = 1.25$$

Since R_o is greater than one, we know that the population must be growing.

TABLE 20.1 Demographic Parameters for a Perennial

Year	Age Class		
	p_x	L_x	m_x
0	0.05	1	0
1	0.10	0.05	0
2	1	0.005	200
3	0	0.005	50

This still does not explain why R_o does not equal R of the simple Malthusian growth model. The reason is that while R_o defines the overall reproductive potential of a female born into a population, it does not take into account the way in which reproduction and survival are distributed. The rate of population increase will be higher in a population in which most of the reproduction is concentrated in the early age classes than in one in which most of the reproduction is concentrated in the old age classes. The population is growing faster when the individuals need less time to mature. For example, Table 20.2 presents the age distribution, total population size, and relative increase in size for several generations based on the demographic parameters given in Table 20.1. Notice that the population does not increase at a steady rate. Actually, it decreases in size every other time period, creating an oscillatory growth pattern. I have extended the data to 18 time periods and have plotted the growth in Figure 20.1. There are two noteworthy characteristics of this pattern of population growth. First, while the growth pattern is oscillatory, it has dampened oscillations that die out over time, which results in a constant rate of increase. Second, once each age class reaches this constant rate of increase, their growth trajectories are parallel. Let us look at these two aspects of demographic growth in more detail.

The dampened oscillatory pattern should remind you of similar general patterns that we observed in predator-prey interactions. We argued that the reason for those oscillations was that the population's growth response was actually being influenced by previous population sizes. In other words there was a time-lag in response. In the predator-prey case this was due to the growth or decrease of the prey population, which caused changes in the population size of the predator population, which in turn resulted in a further response in the prey population. If we were looking at the prey population alone, it was responding to its population size two time periods previous. Thus density response was delayed rather than immediate.

TABLE 20.2 Age Distribution and Relative Population Growth

						Time					
	0	1	2	3	4	5	6	7	8	9	10
N_0	100	14,500	1,850	14,525	5,475	14,987	9,106	16,356	12,853	18,633	16,942
N_1	350	5	725	92	726	274	746	455	818	643	932
N_2	150	35	0	72	9	73	27	75	46	82	64
N_3	200	150	35	0	72	9	73	27	75	46	82
N_{total}	800	14,690	2,610	14,689	6,282	15,343	9,955	16,913	13,792	19,404	18,020
N'/N	—	18.36	0.18	5.63	0.43	2.44	0.65	1.70	0.82	1.41	0.93

Lecture 20   Growth Rate and Equilibrium Age Distribution   205

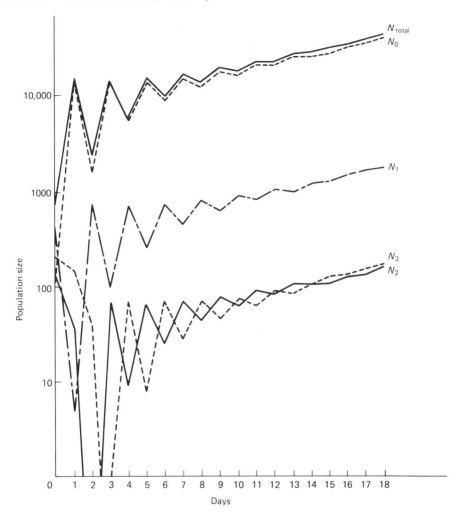

Figure 20.1   Population size of perennial by age class over time.

When there is age structure in a population, the growth is exactly analogous. In the example of the perennial most reproductive output is from the second age class. It takes a full time period for every pulse of newborns, such as in time periods 1, 3, 5, and so on, to filter up to the reproductive class. Thus in our case, the spurt of large population size comes every two periods. If reproduction were delayed longer, the oscillations would have a larger periodicity.

The curves eventually smooth out because each age class eventually makes a constant contribution to growth. In our example the pulses filter up farther into the age structure, so that the older age class produces relatively more exactly at the time when the output of second age class is reduced. Thus the fluctuations of input into the newborn class from one age class is compensated by another. This effect filters up into the older age class, since the survivorship factors are constant. Eventually, the ratios between the sizes of the age classes become constant. When the population has reached this constant age structure, it is said to have a *stable age distribution*.

We are still side-stepping the most important issue of our results. The population sizes in Figure 20.1 are plotted in a logarithmic scale. Eventually, the growth function tends to a straight line. A linear curve on a logarithmic scale is the same as an exponential curve on a linear scale. Therefore, we can write the asymptotic form of the curve as

$$\log N_t = (\log \lambda)t + \log N$$

where $\log \lambda$ is the slope of the curve, N is the original population size, and N_t is the population size after time t. If we transform this to linear scale, we have

$$N_t = \lambda^t N$$

This is exactly analogous to our discrete Malthusian growth equation: λ is equal to R. Thus when a population with a given age structure reaches a stable age distribution, population growth will follow simple exponential growth patterns.

Another point to be understood is that when a population has reached a stable age distribution, each age class remains at a constant proportion of the total population size. Let us call the proportion made up by the age class X, c_X, so that

$$N_X = c_X N$$

The entire population vector is thus

$$\begin{pmatrix} N_0 \\ N_1 \\ \cdot \\ \cdot \\ N_{n-1} \end{pmatrix} = \begin{pmatrix} c_0 N \\ c_1 N \\ \cdot \\ \cdot \\ c_{n-1} N \end{pmatrix}$$

It follows that if the total population grows at a rate λ in one time period, that is

$$N' = \lambda N$$

then any N'_X can also be calculated using the same rate λ, or

$$N'_X = c_X N' = c_X(\lambda N) = \lambda(c_X N) = \lambda N_X$$

Thus the entire age distribution vector is simply

$$\mathbf{N}' = \lambda \mathbf{N}$$

At the same time,

$$\mathbf{N}' = \mathbf{L}\mathbf{N}$$

Lecture 20   Growth Rate and Equilibrium Age Distribution

In other words we can replace the matrix multiplication by a scalar multiplication. Those who are mathematically inclined will realize that λ is an eigenvalue of the Leslie matrix. If you are unfamiliar with linear algebra, do not concern yourself with this definition of λ. The important point to understand is that most populations with age structure will reach a stable age distribution, at which time the total population size as well as the size of each age class will increase at a constant rate λ. The meaning of λ is comparable to the meaning of the parameter R in the simple Malthusian growth model. As in the simple model, we can relate the parameter λ to the intrinsic rate of growth r by the equation

$$\lambda = e^{r\tau}$$

Similarly, remember that λ is defined for a given time period τ which must be considered when calculating the intrinsic rate of increase.

It is rarely sufficiently emphasized that most, but not all, populations will reach a stable age distribution. Among those that do not are populations in which reproductive maturity is delayed, and, once reached, organisms reproduce only once. Biennial plants fit such a description. Again, the longer the delay, the longer the periodicity of oscillations. In such populations several reproducing age classes do not exist to dampen out the oscillations. Pulses of population size are continuously regenerated as the offspring produced during a pulse reach the reproductive class. You should also appreciate that even in those populations in which the stable age structure is achieved, the time it takes to reach it depends on the demographic parameters p_x and m_x.

There is another more direct way of determining the intrinsic rate of growth of a population with age structure if that population is growing exponentially. This method uses the fecundity and survivorship parameters as elements in the *Euler equation*, which I will present shortly. The Euler equation was first derived in the eighteenth century and was introduced in ecology to describe population growth by Lotka in the mid-1920s.

To derive this equation, let us assume that we are dealing with an exponentially growing population. This population is made up of n age classes, with each age class having fecundity and survivorship parameters m_x and L_x, respectively. If the population as a whole is increasing at an exponential rate, then (for the reasons that we discussed earlier) each age class itself increases exponentially at the same rate. Therefore if $N_0(0)$ is the number of newborns at present, then at some later time t the number of newborns will be

$$N_0(t) = N_0(0)e^{rt}$$

Now we can use the same reasoning to reverse this and say that if $N_0(0)$ is the number of newborns alive at present, then at time $(-t)$ in the past the number of newborns must have been

$$N_0(-t) = N_0(0)e^{-rt}$$

I can use this information to determine the number of individuals in all the other age classes. For example, it is logical that the number of individuals in a given age class x must be equal to the number of newborns alive (x) time units ago multiplied by the survivorship probability of reaching age x from age class 0, which is L_x. We can write the number of individuals in age class x at the present time $N_x(0)$:

$$N_x(0) = L_x N_0(0) e^{-rx}$$

To solve for r, I must somehow get rid of $N_0(0)$, but I do not know its value. We can refer back to our discussion in the earlier part of this lecture. $N_0(0)$, the number of newborns in the population, is equal to the sum of the number of offspring of each age class, or

$$N_0(0) = \sum_{x=0}^{n} N_x(0) m_x$$

If I substitute the previous equation into this equation, we get

$$N_0(0) = \sum_{x=0}^{n} L_x N_0(0) e^{-rx} m_x$$

Because the number of newborns at this instant is constant, I can remove it from within the summation.

$$N_0(0) = N_0(0) \sum_{x=0}^{n} L_x m_x e^{-rx}$$

If I divide by $N_0(0)$, I obtain the Euler equation, in which r, the intrinsic rate of increase, is the only unknown.

$$1 = \sum_{x=0}^{n} L_x m_x e^{-rx}$$

I will caution you that this is usually not an easy equation to solve, even when you know the survivorship and fecundity parameters. Often the only way to solve for r is to find it numerically from the equation by an iterative process. First you guess an initial value of r and evaluate the right-hand part of the equation. If the sum is less than 1.0, you guessed too high and you must try a smaller r. If the sum is greater than 1.0, you must try again with a larger r value. By using this method, I determined that r for our perennial is 0.1018. Check my answer to see if it is correct.

By the way, now that we have a value for the intrinsic rate of increase r for our perennial, I can demonstrate an earlier point. We defined the number of female offspring a newborn female is expected to produce as R_o, such that

$$R_o = \sum_{x=0}^{n} L_x m_x$$

R_o for our perennial is 1.25. I cautioned that R_o is not the discrete Malthusian parameter R. In our demographic model λ is equivalent to the Malthusian parameter and is by definition

$$\lambda = e^r$$

Thus in our example, λ is 1.11. This is quite a bit smaller than R_o.

Lecture 20 Growth Rate and Equilibrium Age Distribution

The reasoning behind this result is extremely important, especially when applied to such practical matters as the growth of human populations. For this reason, I will take time to expand our examples and concepts. I mentioned previously that equal R_o's, that is, equal average potential reproductive outputs of a female, can result in vastly different rates of population growth as a result of how the age distribution/life history parameters of a population are distributed. If reproduction is concentrated in the younger age classes, the population will grow more quickly. As a result of this, most individuals of the population will be young. As an example, I will describe another short-lived perennial plant. The demographic parameters of this perennial are given in Table 20.3. Notice that the fecundity per age class is quite a bit lower than

TABLE 20.3 Demographic Parameters of a Short-Lived Perennial

Year	Age Class (x)		
	p_x	L_x	m_x
0	0.1	1	0
1	0.5	0.1	0
2	0.5	0.05	10
3	0	0.025	30

that of the previous plant, but the survivorship coefficients are much higher. Most importantly, the expected replacement rate per female, R_o, is exactly the same as for the previous plant:

$$R_o = 1.25$$

Yet when I estimate the intrinsic rate of growth by means of the Euler equation, it is

$$r = 0.0862$$

This is much less than the 0.1018 value of our original case. One difference is that proportionately more of the reproduction is concentrated in the older age class. Previously, I defined c_j as the proportion of the entire population made up of the age class j, but I did not show you how to calculate it. We now have all the necessary formulae to derive these values. If we are agreed that the number of individuals in an age class j is the number born j time units ago multiplied by the probability of reaching the age of j, or

$$N_j = L_j N_0\, e^{-rj}$$

and the number of individuals in the entire population is simply the sum of all the sizes of the age classes, or

$$N = \sum_{x=0}^{n} N_x = \sum_{x=0}^{n} L_x N_0 e^{-rx}$$

then the proportion c_j can be computed easily:

$$c_j = \frac{N_j}{N} = \frac{L_j N_0 e^{-rj}}{\sum_{x=0}^{n} L_x N_0 e^{-rx}} = \frac{L_j e^{-rj}}{\sum_{x=0}^{n} L_x e^{-rx}}$$

I have used this formula to calculate the stable age distribution of the two plants and have listed the results, as well as the replacement rates and intrinsic rates of increase, in Table 20.4. (Check my results to assure yourself that you understand the concepts.) As you can clearly see, the slower growing population is considerably older. Lecture 21 will develop the practical aspects of this example in more detail.

TABLE 20.4 Stable Age Distribution, Rate of Replacement, and Rate of Increase for Two Short-Lived Perennial Plants

	Species 1	Species 2
c_0	0.950	0.867
c_1	0.043	0.080
c_2	0.004	0.036
c_3	0.003	0.017
R_o	1.25	1.25
r	0.1018	0.0862

Let us quickly review. When a population has an age structure, each age class has a different survivorship and reproductive potential. The fecundity and survivorship parameters together will describe the rate and pattern of population growth. The general patterns of population growth will be characterized by dampened oscillations until the population reaches a stable age structure. At that time, the population as a whole and each age class individually will increase in an exponential fashion. Then, the discrete Malthusian parameter is defined as λ (the first eigenvalue of the Leslie matrix) and is comparable to R of the simple Malthusian growth model. Similarly, the intrinsic rate of increase r is defined by the transformation

$$r = \frac{\ln \lambda}{\tau}$$

where τ is the discrete time unit. In addition r can be derived directly from the Euler equation

$$1 = \sum_{x=0}^{n} L_x m_x e^{-rx}$$

where L_x and m_x are the survivorship and fecundity parameters of each age class, respectively. A population in which the bulk of the reproduction is concentrated in the younger age classes will grow faster than one in which most of the reproduction is delayed. The resultant population itself will be characterized by a younger mean age. This is the case even if the potential number of female offspring a female will produce in her lifetime, R_o, is the same in the two populations.

Our demographic model will be extended to population control problems, questions of environmental impact, and theoretical considerations of the evolutionary forces acting on life history parameters in Lecture 21.

lecture 21
Demography applied

You should now have a good background in density-independent demographic growth models. Lecture 20 ended with a short exposition on the effects of differing age distribution on the growth of two populations, each having an identical replacement rate R_o. The results demonstrated that early reproduction increases the intrinsic rate of increase of the population and results generally in a younger population.

To expand on this idea, I will approach this problem totally intuitively. The problem of initiation time of reproductive activity in the life cycle of an organism can be viewed as a problem involving reproductive overlap between generations. If reproduction begins early in the age of an organism, there is a tendency for total reproduction to bunch up as newly produced offspring rapidly enter into reproductive activity. The earlier the starting age of reproduction, the more overlap of reproducing generations, and thus the more individuals reproducing at one time. This is a self-generating effect, since at any given moment the number of reproducing individuals depends on the number present at a previous time. The outcome is an explosive speeding up of the process.

One interesting application of this result is a method for population control. In organisms such as humans with a long period of essentially constant survivorship (Type I Deevey curve) a delay in the age of first reproduction can reduce the intrinsic rate of increase without the reduction of total lifetime reproductive output of a female. To some extent, the trend of delayed marriage and further delayed childbirth in the United States has contributed greatly to the slowing down of population growth. To be sure, there is a concomitant reduction in births per individual, but this has probably been a less important element in the immediate reduction in r. While encouraging such a sociological change is a complicated problem, other growth controls, such as sterilization or restricted rights of birth, are no less problematic and can lead to a morass of grave ethical and political questions. Certainly, delayed childbirth is the easier, and ultimately probably the more effective, solution.

There are, of course, further conclusions from our simple demographic models that are germane to overpopulation and other problems faced by human societies. As mentioned in Lecture 20, a rapidly growing population is also a young population, with an increasingly greater proportion of the individuals in the newborn and early age classes. Figure 21.1 shows the age distributions

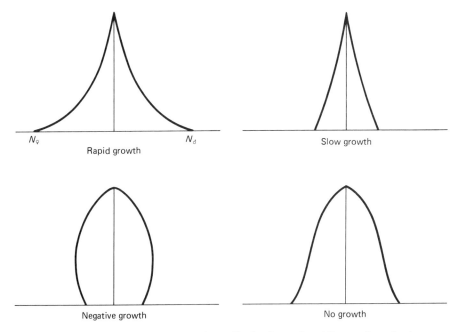

Figure 21.1 Schematic graphs of age distributions of rapidly growing, slowly growing, stable, and declining populations.

of populations with rapid growth, slow growth, no growth, and negative growth. The graphs are traditionally divided into male and female populations. Although survival schedules vary between sexes, the composite graph is nearly symmetric. A rapidly growing population is characterized by a graph that looks like a spruce tree. The base of newborns and young is extremely wide and the sides rapidly become concave as the older age classes are reached. The reason for this shape is the explosive increase in newborns versus the constant decay in cumulative survivorship. I should add that this age distribution configuration does not necessarily imply that the population is rapidly growing. A population with a Type III survivorship curve, such as oysters, that is characterized by an enormous reproductive output but with an equally enormous newborn mortality can have such an age distribution and yet have no population growth.

The following two diagrams show a general trend of a relative narrowing of the base of newborns and an increase in total proportion of the older classes. The slowly growing population has a roughly triangular shape with a flare at the base and the stationary population has a bell-shaped distribution. Depending on the exact survivorship curve, these generalizations may change; for human populations, however, these examples hold up well. In particular the last diagram can be presented with little reservation. Any population that is not sufficiently replacing its newborn class will decrease in size. This is always the case since future older age classes can only come from this newborn class and by necessity will only be some fraction (L_x) of that class. Thus the narrow base will migrate upward over time, decreasing the size of each subsequent age class as it goes.

The implications of these age distributions for human populations can be conceived in terms of the social needs of each age class. The age groups from newborn to 16 or 19 years do not conspicuously contribute to production. Similarly, the age class of 65 and older also consumes more than it produces. These two groups are dependent on the 20 to 65 age groups for many social needs and the majority of their material support. To avoid being overtly controversial, let us take the example of the need for education for the younger age classes. A rapidly growing population has an enormous demand for teachers, classroom material, and educational space. Both the labor force and the resources must come from the relatively small middle age groupings. Inadequate or nonexistent education usually results. Lack of meaningful activities and opportunities may lead to increased violence and crime in later adolescent groups. On a less apocalyptic note, production and social values in such a society are geared to a youth culture. Distinct from this scenario is a stationary or declining population where the production classes are supporting a relatively large, retired age group. This large age group puts a demand on its society to answer its needs in terms of social support and recreational and educational demands. Here production and social values are directed to an older culture.

Another interesting situation arises when a population changes from an expanding to a stationary or decreasing population. The age class that is the most influential in the society moves through time. The United States provides an excellent example of such a situation. Following World War II through approximately 1954, the American society experienced the "baby boom," during which there was a large increase in annual births. In the late 1950s and early 1960s this cohort placed a tremendous demand on the elementary educational system, resulting in an expansion in terms of teachers, physical facilities, and texts. Later in the 1960s and early 1970s this cohort caused a rapid increase in enrollment in colleges and universities. As they finished each system, a vacuum was left in terms of empty elementary schools and a rapid decrease in enrollment in universities and colleges. Concomitant with this movement, the United States developed a strongly influential youth culture from the middle

Lecture 21 Demography Applied

1960s through early 1970s, supported and enhanced by heavy consumer demands. The cohort then moved into the working and producing age classes, creating competition for a limited number of existing jobs. To a large extent, their influence on the culture moved with them, and so the American society became more staid. Finally, this cohort began reproducing themselves (at a later average age than their parents) and a new, large cohort in the young age classes, though smaller than their parents' (the baby boom-boom), is forming.

Although we have touched only briefly on these subjects, it is hoped that an interest in and an understanding of these essentially human demographic problems has developed. Malthus in the eighteenth century recognized these problems, and it is his analysis of the effects of population growth on society, and not of the mechanism of growth alone, for which he is famous.

What are demographic implications for nonhuman populations? In terms of multispecies interactions and the evolution of such interactions these implications can be profound. For example, in Lecture 12 I introduced the problem of the evolution of altruism. In the scenario a male deer in the last year of reproduction defends the herd and thereby decreases his own chances of survival. In that example I couched our thinking of the cost in terms of the loss of the individual fitness versus the benefits of familial fitness. The same question may be posed with a demographic framework in terms of the reduction of population growth or individual reproduction and familial survivorship. We could further limit our discussion to demographic terms by comparing the relative effects of the death of an old, minimally reproducing or nonreproducing individual to the death of a juvenile or young adult.

This leads us to the discussion of the concept, formalized by R. A. Fisher, of the reproductive value of an individual. An individual's *reproductive value* is the number of offspring that individual can be expected to produce in the rest of its life. We can estimate this simply by looking at the number of offspring a whole age class is expected to produce until death or cessation of reproduction divided by the number of individuals in that age class. This then gives a per individual value. You can think of it as the effect of the removal of such an individual on population reproduction. To formalize this concept, consider a population with stable age distribution. We know that the number of offspring the cohort of age x will produce until death will be proportional to

$$\sum_{t=x}^{\infty} e^{-rt} L_t m_t$$

Furthermore, the number of individuals in cohort of age x is proportional to

$$e^{-rx} L_x$$

The reproductive value, which we will denote as V_x/V_0, is then

$$\frac{V_x}{V_0} = \frac{e^{rx}}{L_x} \sum_{t=x}^{\infty} e^{-rt} L_t m_t$$

The notation V_x/V_0 is historically used to represent the ratio of the reproductive values of an individual of age x to that of a newborn. For convenience, the comparison is normalized; that is, V_0 is set to one, and the expression on the right-hand side of the equation can be considered to be V_x.

Of course, the important issue is not the formula or the notation but rather the concept of reproductive value. This concept may be explained in terms of the reproductive value of three individuals of different ages. To make the concept clear, I will take an extreme example of a population with a Type III survivorship curve. The first individual is a newborn whose reproductive value is usually considered to be equal to one for convenience. Before reaching the beginning of reproductive age, it must run the gauntlet of an extremely dangerous and usually fatal part of its life cycle. The second individual is exactly at the beginning of its reproductive period of life. It will obviously produce more individuals on average than a newborn since it has successfully lived through the early stages of its life. The third individual is past its reproductive period. Its reproductive value is necessarily zero, since it will no longer produce any offspring. Let us return to the idea of reproductive value in terms of removing individuals from the population. If there is high mortality in the newborn class, the effect of artificially removing a newborn is minimal since there is a good chance that it would die before reproducing anyway. On the contrary, the most reproductively valuable individuals are those that are just beginning to reproduce, so removing one of these would have the maximal effect. An individual who is no longer breeding has no effect at all on the population growth so that its removal is inconsequential. Of course, if such an older animal is important in the care of others in the population, its removal will have some effect on population growth; our model, however, has no provisions for such a social mechanism. Figure 21.2 is a typical graph of reproductive value as a function of age.

We can use Fisher's concept of reproductive value to make several predictions or observations of evolutionary and ecological importance. I mentioned

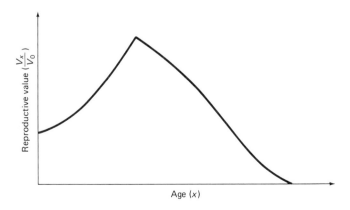

Figure 21.2 Graph of reproductive value as a function of age.

that we can think of fitness and reproductive value as being somewhat alike. This leads to the nonintuitive conclusion that in some sense an individual's fitness will change throughout its life. Genetically controlled characteristics that increase viability or fecundity at ages of high reproductive value will be rapidly selected and established in populations. In counterdistinction, characteristics that improve or even detract from the viability of individuals after their reproductive value has reached zero will be selectively neutral. As a result, traits that are expressed at this point may become randomly fixed. Indeed, if resources are limiting, the death of nonreproducing individuals may even increase the familial or population fitness. It has been suggested that the syndrome of senescence may have developed in this manner.

The whole area of scientific investigation dealing with life history strategies is based on this premise of varying fitness throughout a lifetime. One classical argument deals with the problem of group selection mentioned in passing in Lecture 12. Certain shorebirds are reported to produce fewer offspring per clutch than the number that appears to be the physiological limit. A group selection hypothesis argues that this reproductive restraint occurs to preserve the population from overexpansion, overexploitation of its environment, and, ultimately, extinction. It appears that this selection opposes individual selection, in which it is to the interest of the individual to produce the maximum number of eggs per clutch. An alternative explanation is that this apparent reproductive restraint might not be opposed to individual selection. By reducing the number of eggs per clutch, the survivorship function of the newborns increases during the first year since more resources can be apportioned to each nestling. In addition the total care per year per adult decreases. This decrease, both in energy demand and in predation and accidents that occur during foraging, increases the survivorship during the reproductive period. In the case of birds that have multiple years of reproduction this change in the survivorship increases their reproductive value. Thus the reproductive restraint actually increases rather than decreases individual fitness.

Another interesting implication of demography concerns multispecies interactions. Predators are often observed to concentrate on the consumption of the very young or the old, postreproductive individuals of a prey population. These individuals have the lowest reproductive value and thus their removal may have little effect on the prey population growth. In particular if only postreproductives or weak young are killed, the effects on the growth of the population are close to neutral. In other words on the level of the population, the predator-prey interaction, which we discussed in terms of $(+, -)$ interaction, may often in fact approach a commensal or $(+, 0)$ interaction.

Similar analyses can be incorporated into environmental impact studies. For example, in Lecture 14, I introduced you to the problem of optimal harvesting. In this model individuals were considered to be uniform. Under demographic constraints, we must change some of our assumptions. We cannot concern ourselves with just the number of individuals it is possible to harvest to achieve a maximum sustainable yield, but also with the ages of those individuals that are harvested. The simplest example of this has been mentioned earlier. If we consider ourselves as predators when harvesting, then the impact we make on our exploited population will be best analyzed by using the reproductive values of the harvested individuals. If we only harvest postreproductive individuals, the effect on the population will be minimal. On the other hand, if we capture and collect early reproducing individuals, the effect on the population will be greatest.

Often, analyses of environmental impact compare the effects of two commercially made disturbances to assess the relative contributions to potential environmental deterioration. For example, if you were asked to assess the relative effects of fishing and a power plant disturbance on fish species in a lake or river, you would have to determine fecundity and survivorship schedules of the fish and the changes to those schedules that each form of disturbance might create. Power plants often greatly increase mortality of the newborn class as a result of heat pollution and the physical damage caused by entrainment of eggs and young fish in the intake channels. The heat pollution may also have a secondary effect on fecundity and mating patterns in fish. On the other hand, fishing decreases survivorship in the older age classes, since it is a common practice to catch larger fish. These fish are often the most mature and have high reproductive values. The conclusions drawn from the analysis depend on the life history of each species involved. A species with high first-year mortality, delayed reproduction, and a long reproductive period might be little affected by increased mortality pressures on newborns. Because so many of these individuals are expected to die anyway, the additional death caused by the removal of an egg or a young individual may be of little overall importance. However, removing an individual who is just beginning an extended period of reproduction is more costly. If, however, the species has a Type I or II Deevey survivorship curve and a limited reproductive period early in life, the disturbance by the power plant may be of overriding concern. Any real assessment requires detailed life schedules and knowledge of the absolute levels of impact.

It should be obvious that the demographic model that we have been using has many simplifying assumptions. More realistic considerations, such as density-dependent fecundity and survivorship, can be added, although the mathematical manipulations quickly become complicated. More advanced considerations can take into account the effects of random environmental fluctuations on demographic parameters and the resultant effects on population growth. Needless to say, such considerations are extremely complex. Even by introducing only one random element in the Leslie matrix, the iterative process will result in the randomization of all subsequent age classes. Analyses can then be made on probabilities of amplitudes of population size oscillations, or even on probabilities of extinction.

I have saved the discussion of demography until this chapter because it can be extended to all the models that we have developed to date. The ways

Lecture 21 Demography Applied

in which our general ecological models can be augmented by genetic and evolutionary considerations, especially in light of demography, have been spotlighted. The two fields of ecology and evolutionary genetics are really two parts of one field. They intermesh like the threads of Penelope's weaving. I hope that, unlike that of Penelope, all the weaving that we have accomplished in these lectures will not be unraveled.

Problem solving

For the last problem-solving setting, I thought it appropriate to retire to a sylvan haunt and investigate the demographic growth dynamics of a woodland herbaceous plant. Herbaceous plants are excellent subjects for our purposes for the very reason that they do not fit particularly well into the life cycle pattern that we have discussed. By adjusting our thinking and applying our demographic model to these plant systems, you will get a stronger intuitive understanding of these models.

Before seeking out our study population, let us anticipate the problems and considerations of this system. Our plant, like all angiosperms, has seeds. Each seed has three possible fates which may occur during the year. A seed may germinate, die, or remain dormant. This, then, differs from our previous considerations in which individuals always move from one age class to another or die. The adjustment that we must make for this new contingency is actually small. Consider seeds to be the zero age class. The contribution of each age class to this zero age class is the fecundity parameter. We usually assumed that newborns did not contribute to the next cohort of newborns and m_0 was thus zero. In the case of seeds the probability of remaining dormant is comparable to a nonzero m_0 term. The probability of germinating is then comparable to p_0. These are the only adjustments that we must make in the Euler equation or Leslie matrix to deal with this problem.

As mentioned in Lecture 19, the choice of seeds or seedlings as the first age class is somewhat arbitrary, and depends on the interests of the investigator. Because the investigation of the seed population is filled with technical problems, such as digging out a soil sample, collecting and identifying seeds, and determining whether they are dormant or dead, I have decided for the moment that seeds do not interest me. I will define seedlings as my zero age class.

This leads to two other sticky problems with plants. The first is simply one of definition. Some perennial plants are better divided into size classes rather than age classes. The size of such a plant is often a much better predictor of fecundity and survivorship than age. This alone makes little difference in our demographic models. Yet one correction must be added since a large plant can become smaller or remain the same the next year. This possibility can be correct by incorporating a new survivorship term, p_{ij}, in elements other than the subdiagonal of the Leslie matrix. The second problem is that some plants may reproduce vegetatively. This does not necessarily result in a change in the Leslie matrix since vegetative propagules may be considered as part of the first size class. Vegetative reproduction does become important if one is concerned about the genetic composition of the population.

Let me lead you into the woods. Our site of interest is a 30-meter square clearing along the side of a trail that runs through the woods. The trail used to be a narrow path but a recreation committee recently decided that the woods would provide more general interest if the path were widened for cross-country skiing access. The trail was cut, but luckily has not been overused. In our clearing woodland herbaceous plants, such as violets and foxgloves, have sprung up since the trees were cut and uprooted. The violets are perennials and follow the general life pattern dependent on plant size just described. The foxgloves are biennials that usually flower and die in the second year, although some defer flowering until the third year. I have decided to follow these two populations in time to determine the intricacies of their demography.

I begin by taking a census of these populations for several years. I divide the violets into four size classes: seedlings, nonseedlings with one or two leaves, nonseedlings with three to six leaves, and nonseedlings with seven or more leaves. I can divide the foxgloves into age classes, but I can identify only seedlings in the first year, since I do not know the ages of the mature plants. It appears that no plant lives past its third year, so in the second year I can already estimate yearlings and 2-year-old plants. For both populations, I identify and label each individual plant so that I can track each individual's fate. I also estimate the number of seeds produced by each individual plant. Having collected all these data, how can I estimate the demographic parameters p_x and m_x? How can I arrange these in a Leslie matrix?

> Determining the demographic parameters for the foxgloves is fairly straightforward. The term L_0 is, of course, equal to one. The following two survivorship terms, p_0 and p_1, can be determined by the proportion of seedlings surviving through their first year and yearlings surviving to their second year, respectively, or
>
> $$p_{x-1} = \frac{N_x}{N_{x-1}}$$
>
> where N_x is the number of individuals in the xth age class. I must make a few assumptions to determine the fecundity parameters. I must state

clearly that the fecundity measures that I will be using are not strictly identical to the fecundity terms m_x that I introduced in Lectures 19 through 21. For convenience, I have defined seedlings as my zero age class, yet mature plants do not actually produce seedlings in a given reproductive period (except for cases of vegetative reproduction, with which we are not dealing). However, mature plants do produce seeds that can germinate in the following time period to become seedlings of the zero age class. Therefore, I can define a new fecundity term, F_x, which will be equal to the number of seedlings (zero age class individuals) of the following time period that an individual of the x age class will produce at present. This correction actually simplifies matters since F_x will be comparable to $p_x m_{x-1}$, the terms of the first row of the Leslie matrix. Since we are not marking and following the fate of each individual seed, there is no way of determining whether or not the seeds of the different age classes have different probabilities of germinating. In this case all we can do is assume that every seed has the same probability of germinating. This probability can be estimated by dividing the number of seedlings by the number of seeds available. The number of seedlings that each plant is expected to produce is this ratio times the number of seeds it produces. If S_x is the number of seeds per plant that individuals in age class x produce, T_s is the total number of seeds, and N_0 is the number of seedlings, it follows that

$$F_x = S_x \times \frac{N_0}{T_s}$$

The one problem with this method is that there may be dormant seeds from previous generations that are germinating. This would invalidate our estimation. For simplicity, we will assume that there are no dormant seeds in the soil.

The fecundity measures for the violets can be determined in the same way. Similarly, survivorship parameters can also be determined, although we will have to introduce some new notation. We have used p_x as the probability that an individual of age x will reach age $x + 1$. Now, since we are dealing with size classes, there is a possibility that an individual of size x will remain at size x, or decrease to size $x - 1$ or $x - 2$, or increase to size $x + 1$. To clarify this concept, I will define p_{ij} as the probability that an individual of size class j will be in size class i after one time period.

The two matrices can be arranged as follows. The foxglove Leslie matrix is the same as the general matrix introduced in Lecture 18.

$$\mathbf{L}_{\text{foxglove}} = \begin{vmatrix} F_0 & F_1 & F_2 \\ p_0 & 0 & 0 \\ 0 & p_1 & 0 \end{vmatrix}$$

The violet matrix is not of the general form, but can be written

$$\mathbf{L}_{\text{violet}} = \begin{vmatrix} F_0 & F_1 & F_2 & F_3 \\ p_{10} & p_{11} & p_{12} & p_{13} \\ p_{20} & p_{21} & p_{22} & p_{23} \\ p_{30} & p_{31} & p_{32} & p_{33} \end{vmatrix}$$

After a 7-year study, I have amassed population data on the two species. The foxgloves proved interesting in their fecundity pattern. Most of the plants that reached their second year ($x = 1$) flowered. Only a few did not flower, and these overwintered and flowered in the third ($x = 2$) season. Those that overwintered were larger in the third year than the flowering second-year plants. Perhaps because of this size difference, which may be correlated to the energy available to reproduction, the 2-year-old plants produced a substantially greater number of seeds per plant than the 1-year-old plants. Using the notation just introduced, these figures are

$$S_1 = \frac{20 \text{ seeds}}{\text{plant}}$$

$$S_2 = \frac{35 \text{ seeds}}{\text{plant}}$$

The actual numbers of individuals in each age class are listed in Table VI.1. Using these data, plot population size as a function of time and determine the demographic parameters p_x and m_x. Using these values, determine R_o, R (or λ), and r.

TABLE VI.1 Population Distribution of Foxgloves by Age Class

	Year						
Age Class	1	2	3	4	5	6	7
0	43	53	66	71	86	98	117
1	—	20	24	29	32	39	44
2	—	2	2	2	3	3	3

Let us start by making the assumption that all the year-to-year variation in parameters is nonsignificant and random. (Those familiar with statistical testing can check my assumptions.) For every year I will derive the p_x and F_x values and then simply average them. I will use the formula suggested earlier to determine the fecundity parameters. For the first year of complete data,

$$F_1 = s_1 \times \frac{\text{Number of germinating seeds in the next season}}{\text{Total number of seeds available}}$$

$$= 20 \frac{\text{seeds}}{\text{plant}} \times \frac{66 \text{ seedlings}}{(20 \times 20) + (2 \times 35) \text{ seeds}}$$

$$= 20 \times \frac{66 \text{ seedlings}}{470 \text{ plants}}$$

$$= 2.81 \frac{\text{seedlings}}{\text{plant}}$$

$$F_2 = 35 \frac{\text{seeds}}{\text{plant}} \times \frac{66 \text{ seedlings}}{470 \text{ seeds}}$$

$$= 4.91 \frac{\text{seedlings}}{\text{plant}}$$

I have repeated this for each age class in every year and listed the results and the average fecundity per age class in Table VI.2.

TABLE VI.2 Yearly Age Class Fecundity for Foxgloves (seedlings/plant)

	Year				
	2	3	4	5	6	Average
F_1	2.81	2.58	2.64	2.63	2.64	2.66
F_2	4.91	4.51	4.63	4.60	4.63	4.66

The analysis of the survivorship parameters is straightforward. In the second year I can determine p_0 to be

$$p_0 = \frac{\text{Number of individuals of age class 1 presently}}{\text{Number of individuals of age class 0 one time period previous}}$$

$$p_0 = \frac{20 \text{ plants}}{43 \text{ plants}}$$

$$p_0 = 0.465$$

Similarly, I can begin determining p_1 in the third year as

$$p_1 = \frac{\text{Number of plants of age class 2 presently}}{\text{Number of plants of age class 1 one time period previous}}$$

$$p_1 = \frac{2 \text{ plants}}{20 \text{ plants}}$$

$$p_1 = 0.100$$

Table VI.3 lists all the survivorship values and their averages.

Once I have my average survivorship and fecundity values, determining R_0, R, and r is just a matter of number crunching. R_0 is the average number of foxgloves a foxglove will produce (notice that we are not following females since the foxgloves are hermaphroditic):

Chapter VI   Problem Solving

TABLE VI.3   Yearly Age Class Survivorship for Foxgloves

	\multicolumn{7}{c}{Year}
	2	3	4	5	6	7	Average
p_0	0.465	0.453	0.439	0.451	0.453	0.449	0.452
p_1	—	0.100	0.083	0.103	0.094	0.077	0.091

$$R_0 = \sum L_x F_x$$

$$= \left(0.452 \times 2.66 \frac{\text{seedlings}}{\text{plant}}\right) + \left(0.091 \times 0.452 \times 4.66 \frac{\text{seedlings}}{\text{plant}}\right)$$

$$= 1.394 \frac{\text{seedlings}}{\text{plant}}$$

Since R_0 is greater than one, we can see that the population is growing. The intrinsic rate of increase r will be determined by solving the Euler equation by means of iterative guesses, a method that is not elegant, but works. The equation for the foxgloves is

$$1 = (0.452 \times 2.66 \times e^{-2r}) + (0.091 \times 0.452 \times 4.66 \times e^{-3r})$$

or

$$1 = (1.202 \times e^{-2r}) + (0.192 \times e^{-3r})$$

I will start by setting r equal to 0.1. The right-hand side of the equation is equal to 1.126. Since this is greater than 1, my selection for r is too small. I next try r equal to 0.2 and get a solution of 0.911, implying that 0.2 is too large. By repeating this, I eventually estimate r to be 0.156. I can calculate R or λ directly from this value.

$$R = \lambda = e^r$$
$$R = 1.169$$

I have plotted the population sizes by year in Figure VI.1. The population size is not oscillating over time, as in the example in Lecture 19, but the curve is not perfectly straight. You can notice a blip in the third generation of zero age class that filters up with time to the other age class. Still, the overall smoothness of the curve implies that the population was near its stable age structure from the beginning.

Emboldened by my success with analyzing the foxgloves, I begin analyzing the violet data. I have listed the population distribution by size class in Table VI.4. I noticed immediately that plants seemed to jump size classes as I had expected. For example, in year 4, there are more individuals in size class 3 than in size class 2 in year 3. Obviously, some violets of this size must have come from other size classes. Having marked the individual plants, I know the number that transferred from each size class to another each year. I have listed

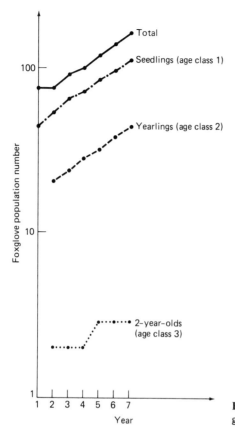

Figure VI.1 Population size of foxgloves by age class plotted over time.

this information in Table VI.5. Last, I recorded the average number of seeds produced by each size class.

TABLE VI.4 Population Distribution of Violets by Size Class

Size Class	Year						
	1	2	3	4	5	6	7
0	54	15	32	52	59	74	96
1	23	22	13	16	27	26	36
2	1	13	13	18	22	35	37
3	0	2	11	12	16	18	30

$$S_1 = \frac{5 \text{ seeds}}{\text{plant}}, \quad S_2 = \frac{9 \text{ seeds}}{\text{plant}}, \quad S_3 = \frac{20 \text{ seeds}}{\text{plant}}$$

Using this information, determine the fecundity and transfer (p_{ij}) parameters. Based on these parameters, what are R_0, $R(\lambda)$, and r? Plot the population size over time.

Chapter VI Problem Solving

TABLE VI.5 Number of Transfers of Plants, N_{ij} from Size Class j to Size Class i

	Year					
	1	2	3	4	5	6
Class 0 (died)	30	9	20	29	41	47
N_{10}	22	6	11	21	15	24
N_{20}	2	0	1	2	3	3
N_{30}	0	0	0	0	0	0
Class 1 (died)	6	7	3	4	3	3
N_{11}	5	4	3	4	8	7
N_{21}	11	9	7	7	14	14
N_{31}	1	2	0	1	2	2
Class 2 (died)	0	0	0	0	0	0
N_{12}	0	3	2	2	3	5
N_{22}	0	4	5	6	9	12
N_{32}	1	7	6	10	9	18
Class 3 (died)	—	0	0	0	0	0
N_{13}	—	0	0	0	0	0
N_{23}	—	0	5	7	9	8
N_{33}	—	2	6	5	7	10

As mentioned previously, the estimation of the fecundity parameters for the violets is exactly like that of the foxgloves. In the first year 124 seeds were produced, of which 15 germinated. In the second year 267 seeds were produced, of which 32 germinated. The first two rows of Table VI.6 list these data for all years. Again, the ratios of germinated seeds to seeds produced are multiplied by the size class specific seed production per plant. For example, the value of F_1 in the first year is

TABLE VI.6 Yearly Seed Production, Seedling Production, and Size Class Fecundity (seedings/plant) for Violets

	Year						
	1	2	3	4	5	6	Average
Total Seeds	124	267	402	482	653	805	—
Seedlings Produced	15	32	52	59	74	96	—
F_1	0.60	0.60	0.65	0.61	0.57	0.60	0.60
F_2	1.09	1.08	1.16	1.10	1.02	1.07	1.09
F_3	—	2.40	2.59	2.45	2.27	2.39	2.42

$$F_1 = S_1 \times \frac{N_0}{T}$$

$$= 5\frac{\text{seeds}}{\text{plant}} \times \frac{15 \text{ seedlings}}{124 \text{ seeds}}$$

$$= 0.60 \frac{\text{seedlings}}{\text{plant}}$$

The fecundity values are listed for all size classes for all years and averaged in Table VI.6.

The estimation of the survivorship (or transfer) parameters, p_{ij}, is also more or less straightforward. I will use the total number of individuals in a size class listed in Table VI.4 and divide it into the number of each size class that transferred to other size classes, which is listed in Table VI.5. For the first year, the transfer parameters are

$$p_{10} = \frac{N_{10}}{N_0} = \frac{22}{54} = 0.41$$

$$p_{20} = \frac{N_{20}}{N_0} = \frac{2}{54} = 0.04$$

$$p_{30} = \frac{N_{30}}{N_0} = \frac{0}{54} = 0.00$$

$$p_{11} = \frac{N_{11}}{N_1} = \frac{5}{23} = 0.22$$

$$p_{21} = \frac{N_{21}}{N_1} = \frac{11}{23} = 0.48$$

$$p_{31} = \frac{N_{31}}{N_1} = \frac{1}{23} = 0.04$$

$$p_{12} = \frac{N_{12}}{N_2} = \frac{0}{1} = 0$$

$$p_{22} = \frac{N_{22}}{N_2} = \frac{0}{1} = 0$$

$$p_{32} = \frac{N_{32}}{N_2} = \frac{1}{1} = 1$$

Notice that there are no transfers to zero size class. We defined that size class as including seedlings only. Seedlings can only come from seeds and not directly from older plants. Therefore, there are no p_{ij} parameters. I have repeated these calculations for all size classes for all years. I have listed these values and their averages in Table VI.7. When calculating the average p_{23} and p_{33} values, I did not use the results of the first years because of the small sample size.

An interesting pattern appears: No plant ever moved from one size class to three size classes away after 1 year. Also, the number moving two size classes away was small. Further, there did not appear to be any

Chapter VI   Problem Solving

TABLE VI.7 Yearly Transfer (Survivorship) Parameters for Violets by Age Class

	Year 1	2	3	4	5	6	Average
p_{10}	0.41	0.40	0.34	0.40	0.25	0.32	0.35
p_{20}	0.04	0.00	0.03	0.04	0.05	0.04	0.03
p_{30}	0.00	0.00	0.00	0.00	0.00	0.00	0.00
p_{11}	0.22	0.18	0.23	0.25	0.30	0.27	0.24
p_{21}	0.48	0.41	0.54	0.44	0.52	0.54	0.49
p_{31}	0.04	0.09	0.00	0.06	0.07	0.08	0.06
p_{12}	0.00	0.23	0.15	0.11	0.14	0.14	0.15
p_{22}	0.00	0.31	0.38	0.33	0.41	0.34	0.35
p_{32}	1.00	0.54	0.46	0.56	0.41	0.51	0.50
p_{13}	—	0.00	0.00	0.00	0.00	0.00	0.00
p_{23}	—	0.00	0.45	0.58	0.56	0.44	0.51
p_{33}	—	1.00	0.54	—	0.44	0.56	0.49

mortality in the third and fourth size classes. Possible accidental deaths do occur to plants in this size class, but the small sample size did not pick this up.

I now have a problem in determining R_o, the number of offspring produced in a lifetime. Since an individual plant may increase and decrease its size indefinitely according to our model, a particular plant may be immortal. In such a case we cannot define R_o exactly as before. There is a comparable value that can be derived, but this goes beyond the basic demography that we have mastered.

On the other hand, the growth parameters r and λ can be determined, although I cannot use the Euler equation as with the foxgloves. Instead, I will use the technique of multiplying the population vector by the Leslie matrix enough times to produce a stable size distribution. At this time, the increase from year to year will be λ (or R) and r will be the natural logarithm of λ. This technique is hardly elegant, but if I have a computer available, it does not entail much work. The Leslie matrix for the violets is

$$L = \begin{vmatrix} 0 & 0.60 & 1.09 & 2.42 \\ 0.35 & 0.24 & 0.15 & 0 \\ 0.03 & 0.49 & 0.35 & 0.51 \\ 0 & 0.06 & 0.50 & 0.49 \end{vmatrix}$$

The value for λ is

$$\lambda = R = 1.2892$$

Taking the natural logarithm, I find

$$r = 0.254$$

I see that the violet population is growing much faster than the foxglove population.

From the graph in Figure VI.2, I see that the population continues to oscillate. There are still no fixed ratios between size classes, as can be detected by noticing intersections of the size class trajectories.

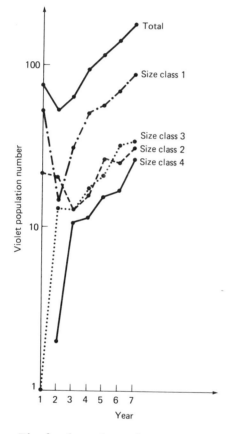

Figure VI.2 Population size of violets by size class plotted over time.

The foxgloves have showy and pretty flowers. By the seventh year, my study patch was attracting a number of observers. Unfortunately, some of these observers felt obliged to take samples. The 2-year-old plants had larger flowers than the yearlings and therefore attracted more predation. What proportion of the population will the 2-year-olds and yearlings eventually reach, barring future disturbances? Will picking a yearling or a 2-year-old have a more deleterious effect on population growth?

The first question may be answered by determining the stable age structure of the population. The proportion made up by an age class i is

$$c_i = \frac{L_i e^{-ri}}{\sum_{x=0} L_x e^{-rt}}$$

Earlier I determined r to be 0.156 for foxgloves. The survivorship parameters are listed in Table VI.3. The proportion made up by the 1-year-old age class is

$$c_1 = \frac{e^{-(0.156)1}(0.452)}{e^{-(0.156)(0)}(1.00) + e^{-(0.156)1}(0.452) + e^{-(0.156)2}(0.091)(0.452)}$$

$$= 0.27$$

The proportion made up by the 2-year-old age class is

$$c_2 = \frac{e^{-(0.156)2}(0.452)(0.091)}{e^{-(0.156)(0)}(1.00) + e^{-(0.156)1}(0.452) + e^{-(0.156)2}(0.091)(0.452)}$$

$$= 0.02$$

There will be more than 10 times the number of yearlings than 2-year-olds.

The effect of picking a 2-year-old or a yearling plant can be determined from its reproductive value, which I defined as

$$\frac{V_x}{V_0} = \frac{e^{rx}}{L_x} \sum_{t=x}^{\infty} e^{-rt} L_t m_t$$

For a yearling plant, the reproductive value is

$$\frac{V_1}{V_0} = \frac{e^{(0.156)1}}{(0.452)} [e^{-(0.156)2}(0.452)(2.66) + e^{-(0.156)2}(0.452)(0.091)(4.66)]$$

$$= 3.023$$

The reproductive value of a 2-year-old plant is

$$\frac{V_2}{V_0} = \frac{e^{(0.156)2}}{(0.452)(0.091)} [e^{-(0.156)2}(0.452)(0.091)(4.66)]$$

$$= 4.66$$

To conclude, picking a 2-year-old plant will have a greater deleterious effect than picking a yearling. Yet, if only one out of every two flowering plants is picked, that is, if p_0 and p_1 are halved, R_0 will be reduced to 0.6486. The population will begin to decline in size. Considering the small size of the patch, all picking should be restricted.

The problem-solving sessions were developed to illustrate the ways in which our theories may be applied to real populations. By chance alone, I began by leading you down a garden path and have finished by taking you into the woods! I sincerely hope that you do not feel that I have also done so figuratively. My goal, even when solving problems, has been to stimulate your thoughts more than to give you formulae.

Homework exercises

There are no long, complicated equations to memorize for the final set of homework exercises. It is important, however, to acquire a basic intuitive understanding of demographic processes, and you must use extreme care in setting up the life tables and the Leslie matrices. The most fundamental problem is misunderstanding the age class subscripts. Remember that zero age class refers to newborns, age class one refers to individuals who have completed their first year of life, and so on. p_x and m_x are the age-specific survivorship and fecundity parameters, respectively. That is, p_1 is the probability that an individual will survive its first year of life and m_1 is the number of offspring it will produce in that period of time. Of course, the time units need not be years; we can use any time period that is convenient and meaningful. Remember also that, as usual, we are only concerned with the female population.

The cumulative age-specific survivorship, L_x, is given by

$$\boxed{L_x = \sum_{i=1}^{x-1} p_i}$$

You can predict the numbers of individuals in each age class in the next time period, $\mathbf{N}'$, from the population age structure vector for this year

$$\mathbf{N} = \begin{pmatrix} N_0 \\ N_1 \\ N_2 \\ \cdot \\ \cdot \\ \cdot \\ N_{n-1} \end{pmatrix}$$

(where n is the total number of age classes) and the Leslie matrix

$$L = \begin{pmatrix} p_0 m_1 & p_1 m_2 & p_2 m_3 & \cdots & p_{n-1} m_n \\ p_0 & & & & \\ & p_1 & & & \\ & & \cdot & & \\ & & & \cdot & \\ & & & & p_{n-1} \end{pmatrix}$$

by matrix multiplication

$$N' = LN$$

Recall the simple rules for matrix multiplication given in Lecture 19.

Most populations with constant age-specific survivorship and fecundity parameters will eventually reach a stable age structure, regardless of the manner in which individuals are initially distributed into age classes. At the stable age equilibrium, the proportions of individuals in each age class will remain constant over time. For populations that have attained the stable age structure, the rate of increase of the entire population, λ, is given by

$$\lambda = \frac{N'}{N}$$

where λ is equivalent to the Malthusian parameter R. We can find r by the usual relationship, $r = \ln \lambda / \tau$, or we can calculate it directly from a knowledge of the demographic parameters by using the Euler equation,

$$1 = \sum_{x=0}^{n-1} L_x m_x e^{-rx}$$

In this case r must be estimated by a process of successive approximation by substituting guesses of its value into the equation.

Finally, the expected reproductive potential of a female born into a population with overlapping generations is

$$R_o = \sum_{x=0}^{} L_x m_x$$

Remember that individuals in two different populations could have equivalent R_o's, but the populations could be growing (or declining) at different rates, depending on the way that reproduction and mortality are apportioned between age classes.

1. Consider a population of beetles infesting a farmer's field. The beetles require 1 month to complete the juvenile (egg, larval, and pupal) stages, during which overall mortality is 90%. Adults live for a maximum of 3 months, and the probability of mortality is constant at 50% for each month. Clutches of 50, 40, and 30 eggs are produced by each female in the first, second, and third months of adult life, respectively (male and female offspring are produced in equal numbers). Complete the life table for this population.

	Month				
Age Class	0	1	2	3	4
p_x				0.0	
L_x	1.0	0.1			
m_x	0.0		20		

2. Suppose that the field is invaded by 20 adult female beetles whose population age structure vector is

$$\mathbf{N} = \begin{pmatrix} 0 \\ 10 \\ 6 \\ 4 \end{pmatrix}$$

Construct the Leslie matrix for this population and calculate the age structure vectors for the next 2 months.

3. What is R_o for this population, calculated from the life table? Use the Euler equation to estimate r.

4. Suppose that 3 months pass before the invading population of beetles is decimated by a pesticide application. The population age structure vectors for the third and fourth months are

$$\mathbf{N}_3 = \begin{pmatrix} 37.5 \\ 14.5 \\ 21.5 \\ 0 \end{pmatrix} \quad \mathbf{N}_4 = \begin{pmatrix} 537.5 \\ 3.75 \\ 7.25 \\ 10.25 \end{pmatrix}$$

Is the population age structure in equilibrium at the time of pesticide application?

5. Consider the following life table:

Chapter VI   Homework Exercises

	Year				
Age Class	0	1	2	3	4
p_x	0.8	0.5	0.2	0.0	0.0
L_x	1.0	0.8	0.4	0.08	0.0
m_x	0	2	2	2	0

How many breeding seasons are required before this population reaches a stable age distribution if the initial age structure is

$$\mathbf{N} = \begin{pmatrix} 20 \\ 10 \\ 5 \\ 0 \end{pmatrix}$$

6. What is λ?
7. What is R_o? Calculate R_o for a population in which fecundity in the age class one is doubled, $m_1 = 4$. Contrast this change with the effect on R_o of a similar increase in the fecundity of the last age class to reproduce, $m_3 = 4$.

lecture 22

Ecological-genetical interactions

To a large extent, it is a historical accident that models of theoretical ecology and population genetics have been developed independently. No doubt important feedbacks operate between evolutionary and ecological processes. The evolution of a species is strongly influenced by the ecological context provided by other species. Likewise, the growth parameters of a population depend on its genetic composition; when a population evolves, these parameters change, too. What right then do we have to consider these two processes independently, as we have done in these lectures and as many authors do? One argument leading to the approximate separation of the two types of models is based on the time scales. In many cases it might be realistic to assume that evolution takes place on a much slower time scale than population dynamics. Ecological variables (population sizes) change much faster than evolutionary variables (gene frequencies). We can therefore "freeze" the gene frequencies while considering population dynamics processes. This should provide a fairly accurate picture, if these processes take place over a short period of time, from an evolutionary point of view. On the other hand, in the coarse evolutionary time scale one can disregard variations in ecological variables, replacing them by their mean values over the large period of time from the ecological point of view.

What if, however, both evolutionary and ecological processes take place on the same time scale? What if changes in both gene frequencies and population sizes are parallel and strongly influence each other? Then we cannot get away with the standard argument anymore. We must consider a model that incorporates both kinds of variables and study population dynamics and evolution explicitly as a joint process. Ecological-genetical models are clearly more complex than models of population genetics or population dynamics taken separately. First, the dimension of the joint system is much higher than that of separate systems. If we look at m populations and distinguish n genetic variables in each population, we will have $m \times n$ variables for the overall system. We have never considered systems with dimensions above two in this book. If you study this or any other subject involving dynamic systems, you will learn that the difficulties of analysis grow drastically with dimension. We know little about high-dimensional systems of differential or difference equations. The mathematics itself is not powerful enough to analyze models of high dimension even if they are formulated in a reasonable way from a biological point of view. The problem of too many parameters, which I mentioned in the Introduction and in Lecture 21 with respect to multispecies models, is even more severe in a model of higher dimension.

The difficulties in developing coevolutionary theory or, in other words, mixed evolutionary-ecological theory, should be apparent. In spite of these difficulties, this area of the theory has recently been the focus of interest of many population biologists, including both theoreticians and experimentalists. Coevolution of plants and plant-pollinating insects is an area where the most striking examples of the wonderful coadaptation of species can be observed. For instance, an innovation that a plant evolves, allowing the new pollinator to act on its flowers, will immediately affect the plant's reproductive ability. Gene frequencies and population sizes will change simultaneously. This is, therefore, a clear evolutionary-ecological process in which one cannot separate the two processes. Interaction between plants and herbivores also provides extreme cases of intricate and fascinating coevolutionary adaptations and is the subject of an active area of research today.

The undesirable and artificial barrier between theories in ecology and evolution is in the process of being slowly destroyed. This destruction is one of the most important areas of research in modern population biology. It is not an established part of the theory, and it will not be discussed in this introductory course. To give you a taste of the kind of problems that are addressed in this interface between evolutionary and ecological theories, let us consider a few simple examples.

The first is density-dependent selection. In our original evolutionary models fitnesses were assumed to be constant over time. We have studied in detail all the possible outcomes of selection for a one-locus, two-allele system. From a population dynamics point of view, the model was simply Malthusian. As long as gene frequencies settle at their equilibrium (polymorphic or monomorphic) values, the population size will grow as a geometric series with the average fitness $\bar{W}$ a Malthusian parameter. Suppose instead of constant fitnesses W_{AA}, W_{Aa}, and W_{aa} we accept the more realistic assumption that they depend on the population size N:

$$W_{AA} = f_{AA}(N)$$
$$W_{Aa} = f_{Aa}(N)$$
$$W_{aa} = f_{aa}(N)$$

Assume also that all three functions decrease with N so that, as in logistic equations, higher population size results in lower average number of viable offspring. The three genotype fitness functions, however, may differ. For instance, one genotype may be the most fit when the population size is low but the superiority may not hold for all values of N. Under crowded conditions another genotype might be the most fit. What will be the outcome of evolution in such a system? This is the problem of density-dependent selection, which is basically a mixed model of population dynamics and our diallelic selection model. The answer to the question depends on the shape of the functions $f_{AA}(N), f_{Aa}(N)$, and $f_{aa}(N)$. In fact it can be shown that it depends only on the relative values of the roots of the three algebraic equations:

$$f_{AA}(N) = 1$$
$$f_{Aa}(N) = 1$$
$$f_{aa}(N) = 1$$

Let us denote the roots of the three equations as K_{AA}, K_{Aa}, and K_{aa}, correspondingly. The notation is not accidental. If the population is purely monomorphic for the A allele, it will grow according to its logisticlike equation and reach the carrying capacity K_{AA} defined by the first equation. In this case the population dynamics model is

$$N' = f_{AA}(N)N$$

and the equilibrium is defined by $f_{AA}(N) = 1$. The system will also be stable as follows from the assumption that all functions are monotonically decreasing. (One more condition is formally necessary for the stability of these equilibria: The absolute values of the derivatives df/dN should not be too large. In discrete time models, such as the one considered here, density dependence that is too strong may lead to oscillations and other kinds of instabilities which we cannot consider here.) The same interpretation is possible for K_{aa}. It is simply the carrying capacity for a population that is purely monomorphic with allele a. The interpretation of K_{Aa} is more difficult. The population cannot be 100% heterozygotic, so let us say K_{Aa} is the root of the equation $f_{Aa}(N) = 1$ and stop at that. We can still refer to K_{Aa} as a carrying capacity for heterozygotes, but we keep in mind that this term is somewhat artificial.

A possible picture of the way functions f_{AA}, f_{Aa}, and f_{aa} may look is given in Figure 22.1. For the low population size, the genotype AA is the most fit, but for the high values of population size the other homozygote is the most fit. The heterozygote in this example stays intermediate for all values of N.

Without proof, in this case the final winner is the allele a, the one with the highest carrying capacity. As in the classical model, one can prove here

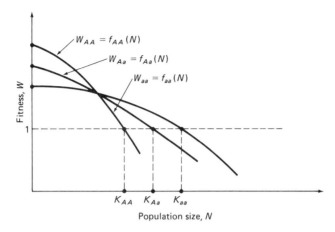

Figure 22.1 Three fitness curves where fitness is a function of population size (density dependent).

that the same four qualitatively different cases are possible. The decisions between the cases are not based on fitnesses, however, but on the comparison of carrying capacities of different genotypes. The evolution will result in polymorphism, for example, if

$$K_{AA} < K_{Aa} > K_{aa}$$

Both of the monomorphic equilibria will be stable if

$$K_{AA} > K_{Aa} < K_{aa}$$

and, depending on the initial condition for both gene frequency and population size, one or the other allele will win.

I do not wish to develop this model in more detail within the framework of this lecture. The lesson is that the conclusions of the classical evolutionary models can be challenged by introducing ecological factors. Of course, if the relationship between fitnesses in our example is uniformly fixed for all N, the old conclusion will apply. If $W_{AA} > W_{Aa}$ for all N, then $K_{AA} > K_{Aa} > K_{aa}$ and the A allele will be fixed in the population. We have no a priori reason to believe, however, that the relationship between fitnesses of different genotypes cannot be reversed under different ecological circumstances. Thus, in general a completely new picture emerges when one takes into account both simultaneous processes.

The other example of an ecological-evolutionary interaction is frequency-dependent selection. This is a model of evolution where fitnesses of an individual genotype may depend on its own frequency and the frequencies of other genotypes. This type of selection is frequently based on interactions with another species. If we present birds (predators) with snails of different colors (prey), the birds will not feed uniformly on all different color morphs. It was shown convincingly by the English evolutionary biologist Bryan Clarke that when the frequency of a certain snail color morph is very low, birds will not consume those snails at the same rate as the dominant color morph. The death rate of snails will therefore depend on the frequency of the given genotype in the population. We can write it mathematically:

$$W_{AA} = f_{AA}(p)$$
$$W_{Aa} = f_{Aa}(p)$$
$$W_{aa} = f_{aa}(p)$$

Here we have used only one argument, p, since in the simplest case of two alleles $q = 1 - p$ and genotypic frequencies are fully defined by p and q based on the Hardy-Weinberg law (panmixia is assumed).

The relationship between the three fitnesses may vary depending on the frequency. A common kind of frequency dependence is the rare type advantage. In other words the fitness of a genotype grows when the genotype is rare. There are a variety of mechanisms leading to the rare type advantage and, therefore, to frequency-dependent selection. When rare, the prey morph is not consumed as often and thus its survivorship increases; when rare, a flower or a male insect mates and reproduces more often. The way our three fitness functions may look is shown in Figure 22.2. In this example the heterozygote is intermediate and W_{AA} is the highest of three fitnesses when the frequency of A (and therefore, AA) is low, whereas the fitness of the genotype aa is highest when the frequency of A is high (and, therefore, the frequencies of a and aa are low).

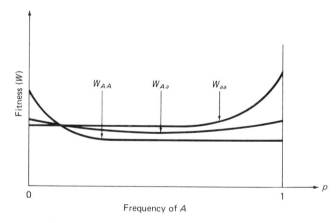

Figure 22.2 Three fitness curves where fitness is a function of allele frequency (frequency dependent).

This type of selection will clearly lead to stable polymorphism. The reason is that if an allele starts to decrease in frequency, it will be helped by the rare type advantage when it becomes rare enough. The frequencies of both alleles will thus stay away from zero. Frequency-dependent selection has, therefore, been proposed as a common cause of polymorphism, more common in the

Lecture 22 Ecological-Genetical Interactions

opinion of some than heterozygote advantage. To what extent this is correct remains to be demonstrated, but the advantage of the rare is without doubt an important evolutionary mechanism. Consider that a new mutant may be favored by selection just because of its initial rarity. A predator will not recognize it as a food item, or a virus adapted to parasitize the common genotype will not be able to attack it. Novelty may have an advantage in a variety of ways. This advantage might be an important mechanism of stimulating evolution. By not letting a rare mutant disappear through pure drift, it gives it a chance to compete with a dominant form and to check fully whether or not it is a viable novelty that will lead to a new evolutionary advance.

As you can see from the two examples, combinations of evolutionary and ecological models may lead to important new qualitative conclusions that could not be obtained on the basis of purely evolutionary or ecological models. They are essentially the result of the "hybrid vigor" between two traditional approaches. Work in this direction has only recently begun; new insights will possibly change our future intuition. Remember that the *counterintuitive* results are usually the most important ones.

Philosophers teach that everything is interrelated and, in addition, all moves and changes. Unfortunately, they are right. Throughout history scientists have tried to find bits of order in this complicated world. Whether or not they have succeeded has depended largely on their ability to find points of view from which to observe and see simple rules. These same rules are obscured if you are looking at them from the wrong angle. Often a seemingly "natural" point of view does not lead anywhere. Ingenious ideas are involved in climbing to a critical "top of the mountain" position from which to view natural phenomena. There are a number of great successes but a much larger, and mostly unrecorded, number of failed attempts. At the beginning of a new work, we always hope that this will lead to a new peak. It takes time to determine whether or not the peak is as great as it seemed at first.

In the introductory lecture I promised to tell you about Galileo's proof which showed why bodies fall to the earth at the same time, regardless of their mass. This example of a marvelous insight is at the same time so intuitive that it is more valuable than any formal proof. Galileo's thought was the following: Consider two bodies, one light and one heavy, and assume that the heavy one falls faster. Glue these two bodies together and see how fast the combined entity will fall. On one hand, the light part should slow down the combination, as compared to the rate at which the heavy part is falling by itself. On the other hand, the combined entity is heavier than either of the two parts and it should fall faster than the heavier part by itself. The way out of the contradiction is to accept that all bodies fall in the same period of time, regardless of their mass. By the way, Galileo himself was so convinced that he was correct that he had his experiments done only to demonstrate to other people that what he believed was really true. He thought that he derived his results from pure thought, without any reference to the experimental evidence. Although his position would not withstand today's philosophical criticism, it shows how much Galileo valued intuition as the source and the driving force of scientific progress. He did throw a few balls from the Tower of Pisa, which showed that his guess was reasonably well supported, but he did it only after he had devised this original, although logically imperfect, proof.

As we grow older and more experienced, we become more skillful in recognizing paths that are clearly wrong. Yet because good ideas are so rare, this skill does not help much in finding the right paths. It follows that the young, energetic, and not very experienced are as likely to encounter good ideas as their more experienced colleagues. I sincerely hope that some of you may be the lucky ones.

Index

A

Adaptive landscape, 109–10
Age class, 194, 196, 220–21
   proportion, c_x, 206, 209–10, 230–31
   social needs, 214
Age distribution graphs, 213–14
Age structure vector, 198–99, 232
Agrostis tenuis, 46
Allee effect, 126, 130, 170
Allele, 9
   substitution under neutrality, 98–99
Altruism, 112, 215
Amensalism, 173–74, 184
Aristotle, 5
Assortative mating, 66
   negative, 66
   positive, 66
Autogamy, 62
Autozygous, 60

B

Baby boom, 214
Balanced polymorphism, 34
Biennial plants, 197, 207, 221
Binomial distribution, 82–83, 107
Birth rate, 120, 122–23, 125–26, 144
Biston betularia, 95

C

Carrying capacity, K, 127–28, 134, 140, 238
Castle, W., 14
Clarke, Bryan, 239
Coevolution, 237
Commensalism, 173–74, 217
Community models, 177–79
Competitive coexistence, 153–59
   stable coexistence, 155–56, 190
   unstable coexistence, 155–56
Competitive exclusion, 145, 147–52, 156, 184–88, 191–92
   principle of, 145
Crow, James, 97
Cystic fibrosis, 45–46

D

Damsel fish, 172–73
D'Ancona, Umberto, 161
Daphnia, 180
Darwin, Charles, 3, 113
Death rate, 120, 122–23, 125–26, 144, 194
Deevey survivorship curves, 194–95, 212, 213, 218
Delayed childbirth, 212–13
Demography, 193–231
Density-dependent coefficient, 126–27, 134, 140

Density-dependent growth, 126, 184
Density-dependent selection, 237–39
Density-independent growth, 126 (see also Malthusian growth)
Desert pupfish, 108
Dinoflagellates, 173
Diptera, 171
Dispersal, 54
DNA, 48–49
 rate of change, 97
 selection on, 111
Dobzhansky, Theodosius, 113
Drosophila, 2, 46, 191–92

E

Ecological efficiency, 162
Ecological matrix, 178
Ecological stability, 176–79
Ecosystem models, 176–79
Ehrlich, Paul, 123
Eldredge, Niles, 113
Electrophoresis, 96
Environmental impact studies, 217–19
Eugenics, 53
Euler equation, 207–8, 210, 225, 233
Evolution, 7–9

F

Familial selection, 111–12, 215
Fecundity, 197–98, 222–24, 227
Fertility, 118, 122–23
Fisher, R. A., 22, 109, 215
Fisher's Fundamental Theorem, 22–28
Fitness, 15–21, 22–27, 29–36, 215, 217
 absolute, 16, 48, 144
 average, 18, 21, 22–27, 117
 components, 16, 21
 marginal, 19, 24
 relative, 17
Frequency, 27, 37–38, 70–71
Frequency-dependent selection, 239–41

G

Galileo, 2, 241–42
Gambler's ruin, 85
Gause, G. F., 145
Gene, 49
Gene flow, 55

Genetic drift, 86–88, 90–95, 103, 105–6
Genotype, 9, 93
Gould, Steven J., 113
Gradualism, 113
Group selection, 112, 217

H

Hardy, G. H., 13, 14
Hardy-Weinberg law, 9–14, 38–39, 42, 44
Harmonic mean, 95
Harper, J. L., 196
Heat pollution, 218
Heterozygosity:
 deficiency, 62–65, 67–69
 expected, 62
 under neutrality, 99–100
Host-parasite system, 172
Hudson Bay Company, 160
Hutchinson, G. E., 146
Hymenoptera, 171

I

Inbred, 59
Inbreeding, 59–69, 94–95
 coefficient of, 59, 64, 65
 due to drift, 94
 F_{IS}, 62, 69, 77–78
 F_{ST}, 67–68, 69, 76, 78
 nonrandom mating, 59–67
 spatial subdivisions, 67–68, 69
Intrinsic rate of growth, 120–23, 124, 127–28, 165, 186
Isocline, 149

K

Kimura, Motoo, 95
Kin selection, 111–12
Kolmogorov, Andrey, 168

L

λ, 206–8, 210, 223–30, 233
L_o, 196
L_x, 194–95, 214, 232
Leslie matrix, 198–200, 201, 221–23, 229, 233
Life history strategies, 217

Limit cycle behavior, 169
Live oaks, 173
Logarithmic growth (see Malthusian growth)
Logistic equation, 127
Logistic growth, 124–31, 134–35, 238
Lotka, Alfred, 147, 161, 207
Lotka-Volterra models, 147–71, 190–91

M

MacArthur, Robert, 128, 169
Macroevolution, 9, 113, 114
Malthus, Thomas, 119, 123, 215
Malthusian growth, 118–23, 124, 132–33
Matrix multiplication, 198
May, Robert, 178–79
Mayr, Ernst, 113
Microevolution, 9, 113, 114
Migration, 54–58, 69
 coefficient, 56
 mutation balance, 56–57
 selection balance, 57–58, 72–73, 77, 103
Molecular clock, 99
Mollusks, 63
Monomorphism, 34
Mutation, 48–53, 69, 97, 105
 base pair substitution, 49
 frameshift, 49
 missense, 49
 neutral, 49
 nonsense, 49
 rates, 50, 52
 selection balance, 52–53, 72–74, 77, 103
Mutualism, 172–73, 174–76
Mytilus edulis, 79
Mytilus galloprovincialus, 79
Myxoma virus in Australian rabbits, 173

N

Nei, Masatoshi, 97
Neutral equilibrium, 13, 163–64
Neutral evolution, 81, 95, 96–101
Neutrality of mutants, 97–98
Newton, Isaac, 3–4
Niche, 145–46
 overlap, 146

O

Oligochaetes, 63
Optimal harvesting, 129–30, 136–39, 186, 217
Overdominance, 32, 72–73, 102

P

Panmixia, 9, 58
Paramecium, 192
Parasitoid systems, 160, 171, 173
Phenotype, 48
Phenylketonuria, 78
Pollination, 62, 173, 237
Polymorphism, 34, 98, 240
Population, 8
 control, 212–13
 vector, 201, 206 (see also Age structure vector)
Predation, 160–71, 172, 188–89, 217, 239
Ptolemaic model, 3–4
Punctuational model of evolution, 113–15

R

r, 120–23, 124, 140, 186, 207, 211, 223–30
R, 118–19, 121–23, 124, 140, 144, 202–3, 206–8, 223–30
r-K strategy, 128
R_o, 202–3, 208–9, 211, 223–30, 233
Refuge for prey, 167
Reproductive overlap, 202–6, 212
Reproductive potential, 202–3
Reproductive restraint, 217
Reproductive value, 215–19, 231
Rosenzweig, Michael, 169
Rotifer, 192

S

Sampling error, 81–89, 104
Sea anemone, 172–73
Selection, 15–21, 22–28, 29–36, 44–45
 artificial, 43
 balancing, 34
 coefficient of, 19, 39, 75
 directional, 34
 disruptive, 34
 migration balance, 57–58, 72–73, 77
 mutation balance, 52–53, 72–74, 77
 stabilizing, 34
 units of, 110–14
Senescence, 217
Shifting balance theory, 109–10
Sickle-cell anemia, 46
Simpson, George Gaylord, 113
Size classes, 221
Spanish moss, 173
Species selection, 113–15

Stability, 31
Stable age distribution, 206, 207, 225
Stable limit cycle, 169
Standard deviation, 86, 87, 88, 107
Stanley, Steven M., 113
Stasis, 113
Stebbins, C. Ledyard, 113
Survivorship, 118, 122–23, 194–95, 224–25, 228–29

T

Time delay, 145, 203–6, 207
Time scales, 234
Transposons, 49, 11
Trophic level, 178, 189
Tschetverikov, S. S., 14
Turnover, 114–15

U

Underdominant, 34, 39, 93

V

Variance, 25, 75–76, 86
Volterra, Vito, 147

W

Wahlund effect, 68
Weinberg, V. W., 13, 14
Wright, Sewall, 109–10

Y

Youth culture, 214–15